SpringerBriefs in Electrical and Computer Engineering

Mohammad A. Matin

Communication Systems
for Electrical Engineers

 Springer

Mohammad A. Matin
Department of Electrical and Electronic
   Engineering
Northern University Bangladesh
Dhaka
Bangladesh

ISSN 2191-8112   ISSN 2191-8120   (electronic)
SpringerBriefs in Electrical and Computer Engineering
ISBN 978-3-319-70128-8   ISBN 978-3-319-70129-5   (eBook)
https://doi.org/10.1007/978-3-319-70129-5

Library of Congress Control Number: 2017956750

Printed on acid-free paper

This Springer imprint is published by Springer Nature
The registered company is Springer International Publishing AG
The registered company address is: Gewerbestrasse 11, 6330 Cham, Switzerland

Contents

Chapter 1
Introduction to Signals, Systems and Communication

1.1 Introduction

Signals and systems are the essential part of electrical engineering that applies mathematical concept in designing electronic devices to perform communication. A signal is the representation of a set of information or data which is processed further by systems to transmit it from the transmitter to the intended receiver. Thus, a system is an entity or quantitative phenomenon of a physical process that converts a set of input signals to another set of signals that is the output. A system could be physical components, as in electrical or mechanical systems or it could be an algorithm that process the input signal and gives an output (Fig. 1.1).

A more complex system can be treated as a set of subsystems. For example, the optical disk reader composed of few subsystems can record and then play back an audio source using a compact disk (CD) storage medium (Fig. 1.2).

A system is called linear when superposition holds. If the response of that linear system is same regardless of the time of the input applied to the system, it is called linear time invariant system (LTI). In this chapter, our focus is on LTI systems as these are good models for a lot of real-life systems. The properties of LTI system guide towards a powerful and effective theory for analyzing their characteristic (Fig. 1.3).

1.2 Signals

Signal is a quantitative representation of a physical process, phenomenon, or event. The categorization of signal leads to the labels analog, digital, continuous-time, discrete-time, periodic, aperiodic, energy and power signals. The terms analog and digital, describe the nature of the signal amplitude whereas the terms

© The Author(s) 2018
M. A. Matin, *Communication Systems for Electrical Engineers*,
SpringerBriefs in Electrical and Computer Engineering,
https://doi.org/10.1007/978-3-319-70129-5_1

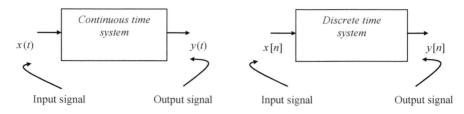

Fig. 1.1 Continuous-time and discrete-time systems

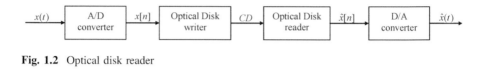

Fig. 1.2 Optical disk reader

Fig. 1.3 Linear time
invariant (LTI) system

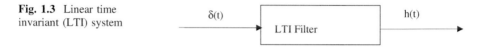

continuous-time and discrete-time, describe the nature of the signal along the time
axis (Fig. 1.4).

An analog signal can be defined as a signal whose amplitude can retain any
value in a continuous range and the analog source is the source that can generate
messages on a continuum. Microphone could be a good example of an analog
source. The output of the microphone is sound and the values of sound signal are
spread over a continuous range. If the signals are not available in the form of
electrical signal, a transducer can be used to have one.

On the other hand, a digital signal is one whose amplitude can retain only a finite
number of values. Digital computer is an example of digital source. The signal in
digital computer can take on only two values (binary signals). However, it is not
necessary to restrict the number of values to two. It can take on any finite number,
M and the signal is called *M-ary* signal.

Continuous-time signal can be denoted as $x(t)$, where t denotes time, which is a
real-valued variable, i.e., $t \in R$. The parenthesis $(\cdot)$ is used to denote a
continuous-time signal whereas discrete-time signal as x[n], where n is an
integer-valued variable denoting the discrete samples of time. We use square
brackets $[\cdot]$ to denote a discrete-time signal.

The signal can also be classified as periodic and aperiodic signals. The signal
$x(t)$ is said to be periodic if the period of the signal satisfies the periodicity condition. Otherwise, it is aperiodic. The periodicity condition is

$$x(t) = x(t+T),$$

where, T is the time period.

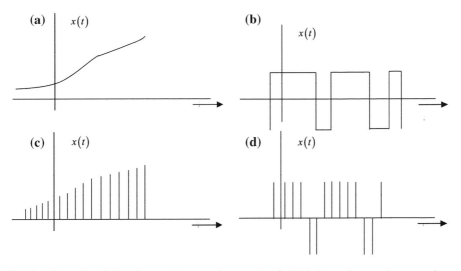

Fig. 1.4 Example of signals **a** analog, continuous time **b** Digital, continuous time **c** analog, discrete time **d** Digital, discrete time

1.2.1 Representation of Signals

The communication is concerned with the transmission and reception of signals. A signal is a means to convey information-it is an electrical voltage or current which varies with time and is used to carry messages or information from one point to another. The signal is function of time. This is not happened always. When an electric charge is distributed over a surface, for instance, the signal is the charge density which is a function of space rather than time. A signal is a function that maps a domain, often time or space, into a range, often a physical measure such as air pressure or light intensity. In this chapter, we deal with signals that are functions of time.

The signals can be denoted by direct mathematical expressions or by using orthogonal series representations such as the Fourier series. The signal of interest can be the voltage as function of time, $v(t)$ or the current as function of time $i(t)$. Often, the same mathematical expression can work either type of waveform of the signal. Thus for generalization, the waveforms can be represented simply as $x(t)$ in case of analysis applied to either voltage or current (Fig. 1.5).

1.2.2 Bandwidth of Signals

In engineering definitions, the bandwidth is taken to be the width of a positive frequency band. In other words, the bandwidth would be $f_2 - f_1$, where, $f_2 \succ f_1 \geq 0$. For baseband waveforms or networks, f_1 is typically to be zero, since the spectrum

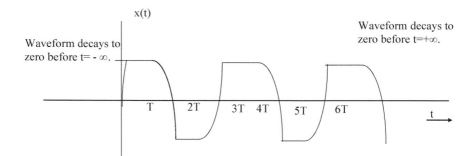

(a) Physical Waveform. .

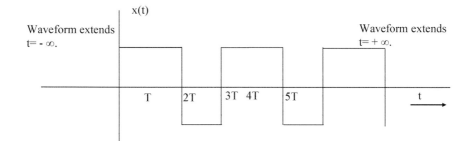

(b) Math Model Waveform

Fig. 1.5 Physical and mathematical waveforms

extends down to dc ($f = 0$). For bandpass signals, $f_1 \succ 0$ and the frequency band $f_1 \prec f \prec f_2$ encompasses the carrier frequency f_c of the signal.

1.2.3   Evaluation of Signal Power

In communication systems, the received average signal power should be adequately large in compare to the average noise power to retrieve the information. The average power is an important parameter that needs to be understood. Let $v(t)$ denotes the voltage across a resistor R, and let $i(t)$ denotes the current flow into the resistor as shown in Fig. 1.6. The instantaneous power associated with the circuit is given as:

$$p(t) = v(t)i(t) \tag{1.1}$$

If $p(t)$ is positive, the instantaneous power will flow into the circuit. In case of negative $p(t)$, power will flow out of the circuit. The average power is

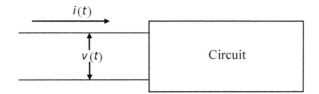

Fig. 1.6 Evaluation of power

$$P = \langle p(t) \rangle = \langle v(t)i(t) \rangle \tag{1.2}$$

For resistive load with unity power factor, the average power is

$$P = \frac{\langle v^2(t) \rangle}{R} = \langle i^2(t) \rangle R = \frac{v_{rms}^2}{R} = I_{rms}^2 R \tag{1.3}$$

The received average signal power should be large enough with respect to noise to recover the transmitted information. The concept of normalized power is often used by communication engineers. In this concept, R is assumed to be 1 Ω, though the actual value is different in the circuit. In the signal to noise ratio calculation, R will automatically efface, so that normalized power can be used to obtain the correct ratio.

1.2.4 Transform of Periodic Signals

A periodic signal can be represented as the integration of an infinite number of harmonic related sinusoids and complex exponentials. The representation of periodic signal $x_p(t)$ with period T_0 is the Fourier series defined as (Fig. 1.7).

$$x_p(t) = \sum_{n=-\infty}^{\infty} c_n e^{j2\pi n f_0 t} \tag{1.4}$$

where, $c_n = \frac{1}{T_0} \int_{T_0} x_p(t) e^{-j2\pi n f_0 t} dt$

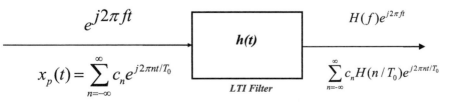

Fig. 1.7 Transformation of periodic signal

The normalized power for the periodic waveform $x_p(t)$, can be expressed as

$$P_w = \prec x_p(t) \succ = \sum_{n=-\infty}^{n=\infty} |c_n|^2, \tag{1.5}$$

where, the $\{c_n\}$ are the complex Fourier coefficients for the waveform and the power spectral density (PSD) of the periodic waveform can be expressed as

$$P(f) = \sum_{n=-\infty}^{n=\infty} |c_n|^2 \delta(f - nf_0), \tag{1.6}$$

where, $T_0 = 1/f_0$ is the period of the waveform. The PSD describes how the power of the signal is distributed throughout the spectrum.

Periodic signals are quit significant in DSP practice. For example, networking over home power lines or a practical issue in audio recording system is eliminating "60 cycle hum," a 60 Hz periodic signal contaminating the audio. Therefore, the tools are required to design a digital filter that would help to get rid of these periodic impurities.

1.2.5   Fourier Transform and Spectra

In the previous section, the representation of analog periodic signals has been discussed. However, to represent and the analysis of finite energy signals, Fourier transform is introduced in this section. Both Fourier series and Fourier transform are very useful in solving engineering problems.

The Fourier transform (FT) of a waveform $x(t)$ is

$$X(f) = F[x(t)] = \int_{-\infty}^{\infty} x(t)e^{-j2\pi nft} dt \tag{1.7}$$

$X(f)$, is called a two-sided spectrum of $x(t)$, as both positive and negative frequency components are present in Eq. (1.7). The FT is applied to get the frequencies in $x(t)$. Evaluate $|X(f)|$, at the arbitrary value of f, say, $f = f_0$. If $|X(f)|$ is not zero, then the frequency f_0 is present in $x(t)$. In general, the FT integral is calculated repeatedly for all possible values of f over the range $-\infty \prec f \prec \infty$ to find all of the frequencies in $x(t)$. The direct calculation of the FT integral would be difficult, therefore the following tables of Fourier transform would be helpful in calculating FT (Tables 1.1 and 1.2).

Table 1.1 Some fourier transform theorems

Operation	Function	Fourier transform
Linearity	$a_1 x_1(t) + a_2 x_2(t)$	$a_1 X_1(f) + a_2 X_2(f)$
Time delay	$x(t - t_d)$	$X(f) e^{-j\omega t_d}$
Scale change	$x(at)$	$\frac{1}{|a|} X\left(\frac{f}{a}\right)$
Duality	$X(t)$	$x(-f)$
Real signal frequency translation	$x(t) \cos(\omega_c t + \theta)$	$\frac{1}{2}\left[e^{j\theta} X(f - f_c) + e^{-j\theta} X(f + f_c)\right]$
Complex signal frequency translation	$x(t) e^{j\omega_c t}$	$X(f - f_c)$
Bandpass signal	$\mathrm{Re}\{x(t) e^{j\omega_c t}\}$	$\frac{1}{2}[X(f - f_c) + X^*(-f - f_c)]$
Differentiation	$\frac{d^n x(t)}{dt^n}$	$(j2\pi f)^n X(f)$
Integration	$\int_{-\infty}^{t} x(\lambda) d\lambda$	$(j2\pi f)^{-1} X(f) + \frac{1}{2} X(0)\delta(f)$
Convolution	$x_1(t) * x_2(t)$	$X_1(f) X_2(f)$
Multiplication	$x_1(t) x_2(t)$	$X_1(f) * X_2(f)$

Table 1.2 Some fourier transform pairs

Function	Time waveform $w(t)$	Spectrum $W(f)$
Rectangular	$\Pi\left(\frac{t}{T}\right)$	$T[Sa(\pi f T)]$
Triangular	$\Lambda\left(\frac{t}{T}\right)$	$T[Sa(\pi f T)]^2$
Unit step	$u(t) = \begin{cases} +1 & t \succ 0 \\ 0, & t \prec 0 \end{cases}$	$\frac{1}{2}\delta(f) + \frac{1}{j2\pi f}$
Constant	1	$\delta(f)$
Sinc	$Sa(2\pi W t)$	$\frac{1}{2W}\Pi\left(\frac{f}{2W}\right)$
Phasor	$e^{j(\omega_0 t + \varphi)}$	$e^{j\varphi}\delta(f - f_0)$
Sinusoid	$Cos(\omega_c t + \varphi)$	$\frac{1}{2} e^{j\varphi}\delta(f - f_c) + \frac{1}{2} e^{-j\varphi}\delta(f + f_c)$
Impulse Train	$\sum_{k=-\infty}^{k=+\infty} \delta(t - kT)$	$f_0 \sum_{n=-\infty}^{n=+\infty} \delta(f - n f_0)$

1.2.6   Signal Shapes in Communication

The waveshapes which are frequently occur in communication problems can be shorted with the special symbols notation such as $\Pi\left(\frac{t}{T}\right)$ for single rectangular pulse, $Sa(\cdot)$ for Sinc type pulse, $\Lambda\left(\frac{t}{T}\right)$ for triangular wave (Fig. 1.8).

1.2.7   Some Useful Signal Operations

There are three useful signal operations: shifting, scaling and inversion which are valid for functions having independent variables. Consider a signal, $x(t)$. If the signal is delayed by T seconds and it is denoted by $y(t)$, then, $y(t) = x(t - T)$.

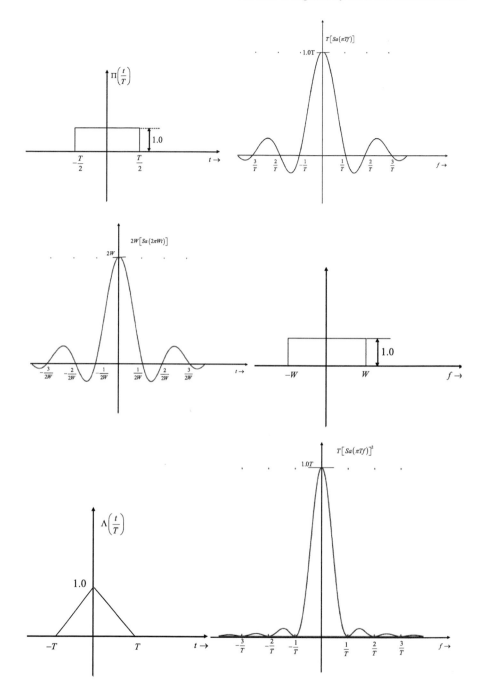

Fig. 1.8 Spectra of rectangular, sinc and triangular pulses

The compression or expansion of a signal in time is known as time scaling. For instance, if the signal $x(t)$ is compressed in time by a factor of 2, then $y(t) = x(2t)$. In case of time inversion, t is replaced with $-t$. The output, $y(t) = x(-t)$.

1.3   System

Mathematically, the system can be modeled as function that maps signals from its input signals—into its output signals. The input and output are both elements of sets of signals; we can call this set of signals in signal space. Thus, a system is an entity that processes a set of input signals to yield another set of output signals. System is a function whose domains and ranges are signal spaces. A system may be made up of physical components or it may be an algorithm that generates output from an input. It is mentioned earlier that our focus of the current chapter is on linear, time-invariant (LTI) system. In this section we will study LTI system.

For a linear, time-invariant (LTI), continuous-time system, the input-output relationship can be expressed using the following equation

$$y(t) = x(t) * h(t), \tag{1.8}$$

where, $x(t)$ is the input, $y(t)$ is the output, and $h(t)$ is the unit impulse response of the linear time-invariant system. If

$$x(t) \Leftrightarrow X(\omega), \quad y(t) \Leftrightarrow Y(\omega), \quad \text{and} \quad h(t) \Leftrightarrow H(\omega)$$

where, $H(\omega)$ the system transfer function, then the Eq. (1.8) is will become

$$Y(\omega) = X(\omega)H(\omega) \tag{1.9}$$

As the input signal $x(t)$ is transmitted through a system, it changes into the output signal $y(t)$. Equation (1.9) shows the type of this change or modification in frequency domain. Here, $X(\omega)$ and $Y(\omega)$ are the spectra of the input and the output of a system, respectively whereas, $H(\omega)$ is the spectral response of the system. The output spectrum, $Y(\omega)$ is given by the input spectrum, $X(\omega)$ multiplied by the spectral response of the system, $H(\omega)$. Equation (1.9) clearly emphasizes the spectral shaping (or modification) of the signal by the system. Equation (1.9) can be written as in polar form:

$$|Y(\omega)|e^{j\theta_y(\omega)} = |X(\omega)||H(\omega)|e^{j[\theta_x(\omega) + [\theta_h(\omega)]]}$$

Therefore,

$$|Y(\omega)| = |X(\omega)||H(\omega)|$$
$$\theta_y(\omega) = \theta_x(\omega) + \theta_h(\omega)$$

During the transmission through channel, the input signal amplitude spectrum $|X(\omega)|$ is modified to $|X(\omega)||H(\omega)|$. Similarly, the input signal phase spectrum $\theta_y(\omega)$ is modified to $\theta_x(\omega) + \theta_h(\omega)$. This means the input signal spectral component of frequency ω is changed in amplitude by a factor $|H(\omega)|$ and is shifted in phase by an angle $\theta_h(\omega)$. Undoubtedly, $|H(\omega)|$ refers to amplitude response and $\theta_h(\omega)$, phase response of the system. During transmission through the system, some frequency components can be boosted up the amplitude, whilst others can be attenuated. The relative phases of the corresponding components are also to be changed. In general, the output waveform will be different from the input waveform. But, our requirement is to get exact replica of the input waveform at the output. Therefore, it is of practical interest to know the characteristics of a system that allows distortionless transmission. For this case, the amplitude response $|H(\omega)|$ must be constant and the phase response $\theta_h(\omega)$ must be a linear function of ω.

1.3.1 Baseband Systems

In baseband systems, the signal is transmitted directly without any modulation. This mode of communication is suitable over a pair of wires, optical fiber, or coaxial cables. It is mainly used in short-haul links. The study of baseband systems is important in two folds. First, many of the basic concepts and parameters encountered in baseband systems are carried over directly to modulated systems. Second, baseband systems serve as a basis against which other systems may be compared.

For a baseband system, the transmitter and the receiver are ideal baseband filters. The low-pass filter at the transmitter limits the input signal spectrum to a given bandwidth. The low-pass filter at the receiver eliminates the out-of-band noise and other channel interference. These filters can also serve an additional purpose, that of preemphasis and deemphasis, which optimizes the signal-to-noise ratio (SNR) at the receiver (Fig. 1.9).

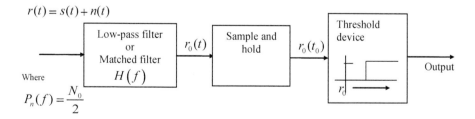

Fig. 1.9 receiver for baseband binary signaling

1.3.2 Passband Systems

The baseband systems have been discussed in the previous section, where signals are transmitted directly without any shift in the frequencies of the signal. However, baseband signals are not feasible for transmission of signal over a radio link or satellites because this would necessitate impracticably large antennas to efficiently radiate the low frequency spectrum of the signal. Therefore, for such case, the signal spectrum must be shifted to a high frequency range. A spectrum shift to higher frequencies is also required to transmit several messages simultaneously by sharing the large bandwidth of the transmission medium. The spectrum of a signal can be shifted to a higher frequency by modulating a high frequency sinusoid by the baseband signal.

1.4 Communication System

A typical communication system consists of transmitter, channel and receiver. Communication can be defined as to exchange information between two parties which can be done face-to-face verbal interaction, over the telephone, through printed materials (letters, newspapers, etc.), or through visual media (television, photographs). Figure 1.10 shows an example of communication system. The basic elements of a communication model are as follows:

The transmitter is required to make the signal suitable for transmission over the channel. For example, if the channel is of fiber-optic cable, the carrier circuit converts the baseband signal to optical signal to propagate through it. If the channel propagates baseband signals, no carrier circuits are needed. The receiver is required to convert the signal in original form from the transmission channel and make suitable for destination.

- Source encoder/Input Device: This device converts the input message into an analog signal or bits called the message signal which is to be transmitted; examples are telephones and personal computers.
- Transmitter: Usually, the data generated by the source encoder are not transmitted directly to the channel. Rather, a transmitter transforms or converts the message signal into a modulated signal in a format which is appropriate for transmission over the channel.

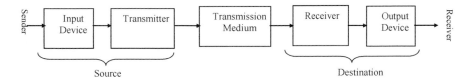

Fig. 1.10 Generic communication system block diagram

- Channel: This can be a single transmission line or a complex network connecting source and destination. It brings up distortion and random noise. Linear channels have output, $y(t) = x(t) * h(t) + n(t)$, where $x(t)$ is the transmitted (modulated) signal, $n(t)$ is the random noise, $h(t)$ is the channel impulse response, and $*$ denotes convolution. Channels may be classified into two categories: wire and wireless. Some examples of wire channels are twisted-pair telephone lines, coaxial cables, waveguides and fiber-optic cables. Some typical wireless channels are air, vacuum, and sea water. In general, the channel medium attenuates the signal so that noise of the channel or the noise introduced by an imperfect receiver causes distortion from the delivered information of the source.
- Receiver: Receiver accepts the distorted or corrupted signal from the channel and extracts the original message signal or bits from the channel output signal.
- Source Decoder/Output Device: Converts the message signal or bits back into the format of the original message.

In communication systems, the received signal waveform is typically the combination of the desired signal containing the information and the undesired signal called noise. The design goal of communication systems is to transmit information to the receiver with highest fidelity as possible while meeting design constraints of allowable transmitted energy, allowable signal bandwidth and cost. In the digital system, the measure of performance of the system is to be the probability of bit error-also called the bit error rate whereas, in analog systems, the performance measure is the signal to noise ratio at the receiver output.

1.4.1 Analog Versus Digital Communication System

Some of the advantages of digital communication system over analog communication system are listed below:

1. Digital communication system is more robust than analog communication system because it can resist the corruption of signal much better in presence of the channel noise as long as the noise is within limits.
2. The greatest advantage of digital communication system over analog communication is the feasibility of regenerative repeaters in digital system. In an analog communication system, a message signal becomes progressively weaker as it travels along the channel (transmission path). If the transmission path is long enough, the channel distortion and noise will add up amply to overwhelm even a digital signal. Therefore, repeater stations along the transmission path are placed at short distances to be able to detect signal pulses before it diminishes. The function of regenerative repeater is to regenerate new, clean pulses.

3. The implementation of digital hardware in digital communication systems is flexible and permits the use of microprocessors, miniprocessors, digital switching, and large scale integrated circuits.
4. Digital signals can be coded to detect and correct error that provide low error rates and high fidelity.
5. In digital system, it is possible to multiplex several digital signals that offer more efficient use of available bandwidth.
6. The storage of digital signal is relatively easy and inexpensive.

1.4.2 Modulation and Demodulation in Communication System

Baseband signals generated from various information sources are not appropriate for long distance transmission over a given channel. These signals are generally further customized to facilitate long distance transmission. Therefore, modulation is necessary. In the modulation, the characteristic of the high frequency carrier is modified based on the message signal which results in different types of analog and digital modulation schemes. At the receiver end, the baseband signal is reconstructed from modulated signal through a reverse process called demodulation. For this purpose, a demodulator circuit is used. The details of modulation and demodulation process are discussed in Chap. 4.

1.4.3 Performance of Communication Systems

The performance of a communication system can be measured in many ways as follows:

• The similarity between the reconstructed signal and the original signal is called signal fidelity. High fidelity means faithful reconstruction of original signal. For analog systems, the performance measure is the output signal-to-noise ratio. The performance measure for a digital system is the bit error rate (BER) of the output signal.
• The required power to transmit the signal. The lower power requirement means longer battery life and less interference.
• The required bandwidth B to transmit the signal. If the system needs less bandwidth, more and more users can be allocated to increasingly crowded RF bands. However, in case of spread spectrum communication system, many users can share the same bandwidth using different user codes.

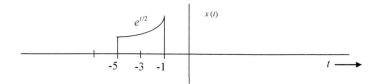

Fig. 1.11 Exponential signal

1.5 Sample Questions

1. Find the $x(t)$ shown in Fig. 1.11, sketch $x(-t)$.
2. Classify the following signals as periodic or aperiodic; for periodic signals, calculate the fundamental period.

$$(a)x(t) = Sin\left(4t + \frac{\pi}{8}\right) \quad (b)x(t) = Sint + Sin\left(2t + \frac{\pi}{8}\right)$$

3. Find Fourier transform of a rectangular pulse of height 0.5, that extend from -1 s to $+1$.

Chapter 2
Transmission Media and Propagation Mechanisms

2.1 Introduction

Signals generated by the source need to be transported to the destination over a communication's channel. A communication channel can be described by its bandwidth, attenuation and the propagation delay. As the signal is attenuated while propagating through the channel, it must be regenerated after a certain distance to maintain signal quality. In general, a communication channel is the physical path between transmitter and receiver and is established through transmission lines in tandem. The transmission links include guided and unguided media such as two-wire lines, co-axial cables, microwave radio, optical fibers and satellites.

2.2 Wired Media: Twisted Pair

Twisted pair cable (Fig. 2.1) is a wire line channel which can be classified as Shielded Twisted Pair (STP) and Unshielded Twisted Pair (UTP). It is least expensive and most widely used wired media. Due to twisting, interference and crosstalk are reduced. The applications of twisted pair cable are

- Data connection especially in PSTN local loop analog links (within an office building, each telephone is also connected to a twisted pair)
- Local area networks supporting personal computers
- In high-speed transmission, although these are quite limited in terms of the number of devices and geographic scope of the network.

The bandwidth of a twisted pair cable varies with the type of wire pair used and its length. Therefore, the use of twisted pair is limited in distance, bandwidth and data rate. For point-to-point analog signaling, a bandwidth of up to 1 MHz is possible for twisted pair and a number of voice channels can be accommodated within this

© The Author(s) 2018
M. A. Matin, *Communication Systems for Electrical Engineers*,
SpringerBriefs in Electrical and Computer Engineering,
https://doi.org/10.1007/978-3-319-70129-5_2

Fig. 2.1 Structure of twisted
pair cable

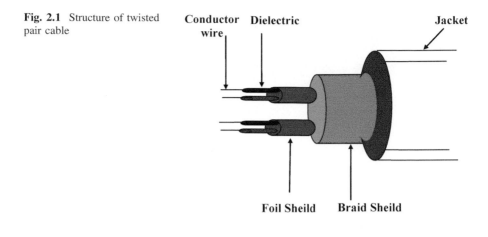

available bandwidth. Coaxial cable, like twisted pair, consists of two conductors, but is constructed differently to permit it to operate over a wider range of frequencies. However, its loss increased drastically with the increased frequency. Thus equalizer is required. Equalization tends to adjust the frequency response. With the introduction of fiber optic cable, with its much greater bandwidth and comparatively flat frequency response, the coaxial cable becomes obsolete though it is also extensively used in cable television plants, especially in the "last mile".

2.3   Wired Media: Optical Fiber

An optical fiber is a dielectric waveguide capable of transporting light waves. The optical fiber media is used for high speed data transmission. The primary material for the construction of optical fiber is nonmetallic and non-conductive. There is no electromagnetic radiation from the optical fibers due to waveguide transmission with optical flux contained within the fiber waveguide (Fig. 2.2).

Optical fiber is a long, thin strand of very pure glass about the diameter of 2–125 μm. Optical fibers are assembled together called optical cables and used to transmit light signals over long distances. The propagation through optical fibers is entirely based on the principle of total internal reflection. This is explained in the following picture.

2.3.1   Structure of an Fiber Optic Cable

Typical optical fibers consist of the core, cladding and buffer or coating. The core is the innermost part of the fiber, which guides light. It is generally made of glass. The core is surrounded with cladding. The cladding has optical properties different from

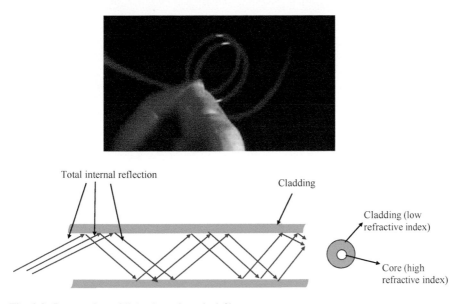

Fig. 2.2 Propagation of light through optical fiber

those of the core, light in the core strikes the interface between the core and cladding and confine light that would otherwise escape the core. The outermost part surrounding the cladding is the jacket.

For the most common optical glass fiber include 1550 nm single mode fibers and 850 or 1300 nm multimode fibers. Multimode fiber has larger core diameter than the single mode fiber and the three most common core diameter sizes are 9 µm (single mode), and 50 or 62.5 µm (multimode). The most common cladding diameter is 125 µm. The material of buffer coating is soft or hard plastic such as acrylic, nylon and with diameter ranges from 250 to 900 µm. The buffer coating provides mechanical protection and has bending flexibility for the fiber (Fig. 2.3).

2.3.2 Propagation Modes of Fiber Optic Cable

The light waves are propagating through optical fiber in different modes. Modes define the way the light waves travel across the fiber. The waves can have the same mode but have different frequencies. It depends on the variation in refractive index that shapes the core. The variations in refractive index create boundary conditions that dictate traveling of light waves through the fiber, like the walls of a tunnel effect that create sounds echo inside.

The variation of material composition of the core gives rise to the two commonly used fiber types. In the first type, the refractive index of the core is constant throughout and undergoes an abrupt change at the cladding boundary. This is called

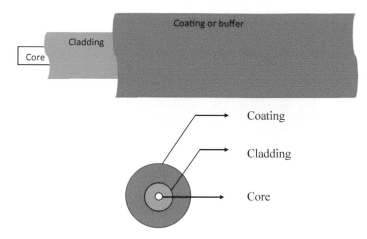

Fig. 2.3 Structure of optical fiber

step-index fiber. In the second type, the refractive index of the core is varied as a function of radial distance from the center of the fiber. This type is called graded-index fiber. Both the step-index and graded-index fibers can be further classified into single mode and multimode.

Fibers that support many propagation paths or transverse modes are called multimode fibers. It has a wider core diameter. Light waves travel at slightly different velocity in a multimode fiber. Therefore, pulse arrives at the end of fiber at slightly different times which causes pulse to spread out in time. On the other hand, some fibers have narrow core diameter that can support only one mode. Light wave travels through the center of the core as a straight line. These fibers are single mode fibers. This is illustrated in the following picture (Fig. 2.4).

2.3.3 Calculation of Number of Modes in a Fiber

Modes are sometimes expressed by numbers. Single mode fibers support only the lowest-order mode, assigned the number 0 (zero) and multimode fibers support higher-order modes. The number of modes that can propagate through a fiber depends on the fiber's numerical aperture (or acceptance angle) as well as on the diameter of its core and the wavelength of the light. For a step-index multimode fiber, the number of such modes, Nm, is approximated by

$$N_m = 0.5\left(\frac{\pi D \times NA}{\lambda}\right)^2 \tag{2.1}$$

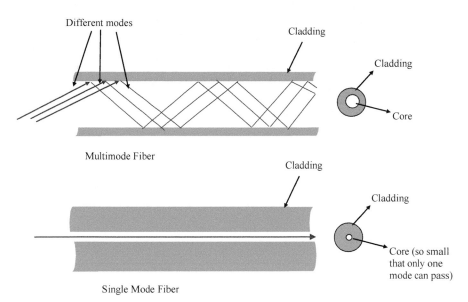

Fig. 2.4 Single mode and multimode Fiber

where, D is the core diameter, λ is the operating wavelength NA is the numerical aperture (or acceptance angle). The above equation is only an approximation and does not work for fibers carrying only a few modes.

Problem 2.1

A step-index fiber has a normalized frequency, V = 26.6 at a 1300 nm wavelength. If the core radius is 25 μm, let us find the numerical aperture and total number of modes entering the fiber.

Solution

$V = \frac{\pi D}{\lambda} NA$

$$NA = \frac{V \times \lambda}{\pi D} = \frac{26.6 \times 1300 \times 10^{-3}}{3.1416 \times 2 \times 25}$$

$NA = 0.220$

2.3.4   Optical Fiber Index Profile

The Index profile is the distribution of refractive indices across the core and the cladding of the fiber. In the case of step index profile, the core has one uniform refractive index and a sharp decrease in refractive index at the core-cladding interface so that the cladding is of lower refractive index. For a graded index profile,

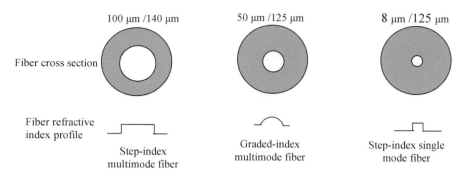

Fig. 2.5 Optical fiber index profile

refractive index varies steadily as a function of radial distance from the optical axis of the fiber. The graded-index profiles include power-law index profiles and parabolic index profiles. The following figure shows some common types of index profiles for single mode and multimode fibers (Fig. 2.5).

2.3.5 Optical Fiber's Numerical Aperture (NA)

Numerical aperture is a measure of the acceptance angle of fiber. The propagation of light through multimode optical fiber occurs if enters the fiber within acceptance angle of the fiber. For step-index multimode fiber, the acceptance angle is determined only by the indices of refraction of core, the cladding and the medium:

$$NA = n \sin \theta_{\max} = \sqrt{n_f^2 - n_c^2} \qquad (2.2)$$

where, n denotes the refractive index of the medium, n_f is the refractive index of the fiber core, n_c is the refractive index of the cladding (Fig. 2.6).

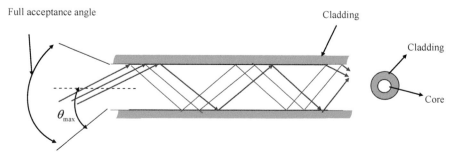

Fig. 2.6 Numerical aperture

2.4 Wireless Media

Wireless communication has a rapid progress in the last decade, and many types of wireless systems have flourished, and often later on, these are vanished. For example, television transmission, in its early days, was broadcast by wireless radio transmitters, which are being swapped by cable transmission. Likewise, the telephony system, point-to-point microwave circuits are being restored by optical fiber. In the first example, wireless technology happened to obsolete due to wired distributed network; in the second case, wired technology (optical fiber) take over the older technology. The opposite example is true in today's telephony, where wireless (cellular) technology is replacing to some extent the use of the wired telephone network. The point of these examples is that the modern communication has given us some options in different situations to select whether wireless or wire technologies and the choice often changes when new technologies become available.

There are different propagation phenomena of wireless communication such as reflection, diffraction and scattering that create the problem challenging and exciting which is not as significant as in wireline communication. The first phenomenon is fading which is due to the rapid fluctuations of signal strengths. The effect of small-scale multipath fading is attenuation while for larger-scale fading are path losses and shadowing. The wireless users communicate over free space and the interference is significant between transmitters communicating with a common receiver (e.g., uplink of a cellular system), between signals from a single transmitter to multiple receivers (e.g., downlink of a cellular system), or between different transmitter–receiver pairs (e.g., interference between users in different cells).

The propagation at LF band is due to ground wave which provides stable transmission over distances up to about 1500 km. This band is used for long-wave sound broadcasting. In the MF and HF bands, sky wave predominates and these bands are used for sound broadcasting and long distance communication to ships and aircraft.

2.5 Transmission Impairments

In any communications system, the received signal can differ from the transmitted signal due to various transmission impairments. Transmission impairments cause the signal quality degradation for analog signals, and bit errors is introduced for digital signals, for example a binary 1 is converted into a binary 0 or vice versa.

Attenuation

Attenuation is the falls off signal strength with the distance during propagation over the transmission medium. For guided media, this is usually exponential and is normally expressed as a constant number of decibels per unit distance. For unguided media, attenuation is a more complex function of distance. In microwave (and radio frequencies), the loss can be expressed as:

$$L = 10 \log \left(\frac{4\pi d}{\lambda} \right)^2 dB, \qquad (2.3)$$

where, d is the distance and λ is the wavelength, in the same units. Thus loss varies as the square of the distance.

To overcome attenuation, the transmitted signal strength need to be increased so that the received signal have sufficient strength to detect the signal and the signal must maintain a level sufficiently higher than the noise to be received without error. The attenuation problem is dealt with the use of amplifiers or repeaters. Techniques are also available for equalizing attenuation across a band of frequencies. For voice-grade telephone lines, loading coils are used that change the electrical properties of the line which smooth out attenuation effects. Another approach is to use amplifiers that amplify high frequencies more than lower frequencies.

Delay Distortion

Delay distortion happens due to the signal velocity through a guided medium. For a bandlimited signal, the velocity is likely to be highest near the center frequency and decrease towards the edges of the band. Thus different frequency components of a signal will arrive at the receiver from different directions with different times, creating phase shifts between the different frequencies component. Delay distortion is critical for digital data, because some of the signal components of one bit position will spread out into other bit positions, causing intersymbol interference. This is a major limitation to achieve maximum bit rate over a transmission channel [1].

Noise

For any communication system, the received signal consists of the transmitted signal which is modified by the transmission impairments, plus additional unwanted signal, referred to as noise, that are added anywhere between transmission and reception. Noise is a major issue in communication system performance. Thermal noise cannot be purged and therefore places an upper bound on communication system performance; and. is mainly important for satellite system. Thermal noise is generated from the thermal agitation of electrons in all electronic devices and transmission media and is a function of temperature. Thermal noise is evenly spread across the bandwidths usually used in communication systems and hence is often referred to as white noise.

Intermodulation noise is generated due to transmission of signals at different frequencies sharing the same transmission medium. Due to the intermodulation noise, the signals are produced at a frequency that is the sum or difference of the two original frequencies or multiples of those frequencies, thus probably create interference with services at these frequencies. It is produced by nonlinearities in the transmitter, receiver, and/or intervening transmission medium.

2.6 Data Transmission

The successful transmission of data mainly depends on two factors: the quality of
the signal being transmitted and the characteristics of the transmission medium.
There are a variety of impairments that can distort or corrupt a signal. The chief
impairments are attenuation, delay distortion, noise. For digital transmission these
impairments limit the data rate. Data rate is the maximum rate at which data can be
transmitted over a given communication channel, under given conditions. There are
four parameters that are related to one another—data rate, bandwidth, noise and bit
error rate and determine channel capacity.

- Data rate, in bits per second (bps), at which data can be transmitted
- Bandwidth depends on the nature of the transmission medium, expressed in
  cycles per second, or Hertz
- Noise, average level of noise exists over the communications bandwidth
- Error rate is the rate of error, at which errors occur, where the error was the reception
  of 1 when a 0 was transmitted or the reception of a 0 when a 1 was transmitted.

All the transmission channels are of limited bandwidth due to the physical properties of the transmission channel or intentionally limited the transmission bandwidth to prevent interference from other sources. The ultimate goal is to make an
efficient use of available bandwidth as much as possible. For digital data, this means
that we would like to get as high data rate as possible at a particular error rate for a
given bandwidth. The main constraint on achieving this high data rate is noise.

2.6.1 Nyquist Information Capacity

Nyquist showed that the theoretical minimum bandwidth needed for the baseband
transmission of R_s symbols per second without ISI is $R_s/2$ Hz. If the transmitted
signals are binary (two voltage levels), then the data rate that can be supported by B
Hz is 2B bps. However, signals with more than two levels (i.e. multi-level signaling) can be used; that is, each signal element can be represented by more than
one bit. For example, if eight possible voltage levels are used as signals, then each
signal element can be represented by using three bits. With multilevel signaling, the
Nyquist formulation becomes

$$r_b = 2B_T \log_2(L), \qquad (2.4)$$

where, L is the number of discrete signal or voltage levels. Thus, for a given
bandwidth, the data rate can be increased by increasing the number of voltage
levels. However, this imposes an augmented burden on the receiver, as it must
differentiate one of L possible signal elements. Noise and other impairments on the
transmission channel will limit the practical value of L. However, there is absolute

maximum of information capacity that can be transmitted in a channel. This is called as (Shannon's) channel capacity, which is given in the next section.

2.6.2 Shannon Capacity

There is a relationship among data rate, noise, and error rate. Due to the presence of noise, one or more bits can be corrupted i.e. efficient transmission is controlled by noise. If the data rate is increased, then the bits become "shorter" so that more bits are affected by a given pattern of noise. Mathematician Claude Shannon developed a formula relating these.

Shannon's result is that the maximum channel capacity, in bits per second, complies with the equation shown below. C is the capacity of the channel in bits per second and B is the bandwidth of the channel in Hertz. For a channel with received signal power, S additive white noise with received noise power, N. The Shannon formula represents the theoretical maximum that can be achieved. In practice, however, only much lower rates are achieved, in part because formula only assumes white noise (thermal noise).

$$C = B \log_2(1 + S/N) \text{ bps} \tag{2.5}$$

For a certain level of noise power, the ability to receive data correctly in the presence of noise depends on the signal strength. The signal-to-noise ratio (SNR, or S/N) is an important parameter that sets an upper bound on the achievable data rate. Typically, this value of S/N ratio is measured at the receiver, because it is at this point that an attempt is made to process the signal to recover the data. For the convenience, this ratio is often reported in decibels. This expresses the amount, in decibels, that the intended signal exceeds the noise level. A high SNR will mean a high-quality signal and a low number of required intermediate repeaters.

However, Shannon theory does not tell you how to design real communication systems. Shannon theory predicted a maximum modem speed of 32 Kbps in 1949. Today, we have 56 Kbps modems.

Problem 2.2
A mobile communication system uses a radio channel of bandwidth 30 kHz. with an intended capacity of 180 Kbps. What signal to noise ratio is required to achieve this capacity (considering AWGN)?

Solution

Given,

Channel Capacity, C = 180 Kbps = 180 × 10³ bps
Channel Bandwidth, B = 30 kHz = 30 × 10³ Hz
Find SNR = ?

According to Sahnnon's Channel Capacity,
The theoretical bit rate, $C = B \times \log_2(1 + SNR)$
$180 \times 10^3 = 30 \times 10^3 \log_2(1 + SNR)$
$1 + SNR = 2^6$
$SNR = 2^6 - 1$
$SNR = 63$
$SNR_{dB} = 18.06$ dB (ans).

2.7 Wireless Propagation

A radio signal radiated from antenna propagates through atmosphere in a path between transmitter and receiver. There are two different waves (ground wave and sky wave) of carrying messages from the transmitter to a receiver. The ground wave is used for short-range communications with high frequencies and at low power, and for long-range communications at low frequencies and with very high power. Daytime reception from most commercial radio stations operating in the medium frequency (MF) band is carried by the ground wave.

The sky wave is used for long-range, high frequency communications. Due to varying ionospheric conditions, the daylight frequencies for sky wave propagation are somewhat higher thin at night.

2.7.1 Ground Wave Propagation

The ground wave consists of two separate components called the space wave and the surface wave. The classification of ground wave components lies in whether the component wave is traveling along the surface of the earth or over the surface. One factor is that the electromagnetic wave induces a current in the earth's surface which slows the wave front near the earth, causing the wavefront tilt downward and hence follows the earth's curvature. The other factor is diffraction due to the presence of obstacles (Fig. 2.7).

2.7.2 Sky Wave Propagation

The sky wave, often referred to as the ionospheric wave, is radiated in an upward direction and it returns to the earth at some distant place. The return point is due to refraction from the ionosphere. The sky wave travels through a number of hops, bouncing back and forth between the ionosphere and the earth's surface. This form of propagation is relatively unaffected by the earth's surface and is capable the HF Band (3–30 MHz) is used in sky wave propagation (Fig. 2.8).

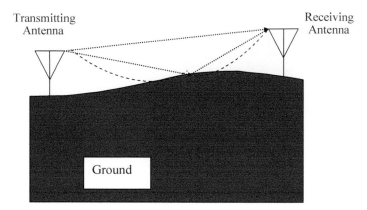

Fig. 2.7 Schematic diagram of ground wave propagation

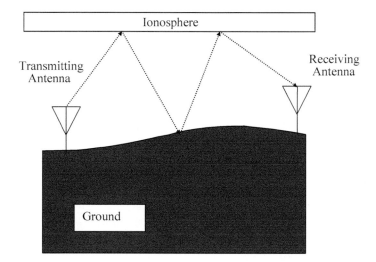

Fig. 2.8 Schematic diagram of sky wave

The critical frequency is the limiting frequency at or below which a radio wave is reflected by an ionospheric layer. Radio waves which are higher than the critical frequency of a given ionized layer will pass through the layer and be lost unless refracted by an upper, more densely ionized layer operating with a higher critical frequency. Radio waves of frequencies lower than the critical will be refracted back to the earth. The critical frequency can be determined using the following formulae:

Critical Frequency, $f_c = 9\sqrt{N}$; where, N is equal to the number of electrons per cubic meter.

Problem 2.3
Calculate the critical frequency if the maximum electron density (N_{max}) of the layer used is 4×10^{11} electrons/m^3.

Solution

$$F_{crit} = 9\sqrt{N_{max}} = 9\sqrt{4 \times 10^{11}} = 5.693\,\text{MHz (Ans)}$$

2.8 Propagation Mechanism

Electromagnetic waves propagate through environments where they are reflected, scattered, and diffracted by walls, terrain, buildings, and other objects. The ultimate details of this propagation can be obtained by solving Maxwell's equations with boundary conditions that express the physical characteristics of these obstructing objects. The reflection, diffraction and scattering mechanisms are briefly explained in this section. Reflection occurs when a propagating electromagnetic wave strikes upon an object which has large dimension in compare to the wavelength of the propagating wave. Reflections occur from the surface of the earth and from buildings and walls.

Diffraction occurs due to the presence of obstacle in between the transmitter and receiver that has large scale irregularities (edges) such as hills, trees, or buildings. It allows radio signals to propagate behind the obstacles. The amount of loss depends on the shape of terrain elevations, the electrical characteristics of the ground, the frequency of operation and the extent of Fresnel zone obstruction between the transmitter and receiver.

Scattering occurs when the medium through which the wave travels consists of objects with dimensions that are small compared to the wavelength, and where the number of obstacles per unit volume is large. Scattered waves are produced by rough surfaces, small objects, or by other irregularities in the channel.

In wireless communication, a line-of-sight (LOS) path seldom exist due to the presence of different obstacles between transmitter and receiver. Even if, there are no other sources of attenuation in case of LOS transmission, the transmitted signal can realize attenuation as the signal is being dispersed over distance. This form of attenuation is called free space loss. To calculate the free space loss, consider a signal transmitted through free space to a receiver located at distance d from the transmitter.

$$P_r(d) = \frac{P_t G_t G_r \lambda^2}{(4\pi)^2 d^2 L}, \tag{2.6}$$

$$Path\,loss = 10\log_{10}\frac{P_t}{P_r} = -10\log_{10}\left[\frac{G_t G_r \lambda^2}{(4\pi)^2 d^2 L}\right]$$

where, $P_r(d)$ is the received signal power at distance d from the transmitter, P_t is transmitted power, G_t and G_r are transmitter and receiver antenna gains respectively, λ is the wavelength and L is the system loss factor. The received power is proportional to $1/d^2$ or d^{-2} but in real life measurement the received power signal level tends to follow a d^{-n} curve, where n is the number typically between 2 and 6, referred to as the *path loss exponent*.

2.9 Sample Questions

1. What are the primary functions of transmission systems?
2. What is attenuation?
3. Define channel capacity.
4. What key factors affect channel capacity?
5. A digital signaling system is required to operate at 9600 bps.
6. If a signal element encodes a 4-bit word, what is the minimum required bandwidth of the channel?
7. Repeat question (6) for the case of 16 bit words.
8. Define channel capacity. A channel with an intended capacity of 20 Mbps and the bandwidth of 3 MHz is given. Calculate the required signal to noise ratio to achieve this capacity. Assume white thermal noise.
9. What is the difference between guided media and unguided media?
10. What is the difference between shielded and unshielded twisted pair cable?
11. Describe the components of optical fiber cable.
12. Distinguish between the terms ground wave and sky wave used in radio wave propagation.
13. What is the difference between diffraction and scattering?

Reference

1. W. Stallings, *Data and Computer Communication*, 10th edn. (Pearson, 2013)

Chapter 3
Speech Digitization and Service Integration

3.1 Introduction

The acceptable level of speech transmission is obtained in the frequency range of 400–3400 Hz and our ear is sensible to frequencies that are around 3 kHz. In analog speech transmission, one of the shortcomings is noise and interference which is more significant during pauses. In digital transmission, speech and speech pauses are encoded with data pattern and transmitted at a constant power level which ensures better quality, provides higher capacity and deals with longer distance. The main drawback of digital transmission is the requirements of greater bandwidth. Nonetheless, the benefits of digital transmission outweigh the bandwidth consideration.

3.2 Speech Digitization

The range of frequencies in analog speech signals is extended up to a maximum of 4 kHz. This speech signal can be dealt as a sequence of bits or data instead of continuous-time analog waveforms. This has several benefits including the integration of voice and data services. Moreover, data can be switched and transmitted over more reliable and convenient network.

For voice digitization in the PSTN, the sampling rate is 8 kHz according to the Nyquist sampling theorem. Let V (volts) is the maximum amplitude of the input speech. Each speech sample is approximated to fall in one of M levels within the dynamic range of $[-V, V]$. In PSTN, the number of quantization levels, M is 256 and V is of the order of 1 V. This allows precise quantization of the input speech. Each quantized sample is encoded using 8-bits code word. This generates a bit rate of 64 Kbps if applies pulse code modulation.

© The Author(s) 2018
M. A. Matin, *Communication Systems for Electrical Engineers*,
SpringerBriefs in Electrical and Computer Engineering,
https://doi.org/10.1007/978-3-319-70129-5_3

As voice signal has non-uniform distribution, non-uniform quantization technique is applied to maximize signal to noise ratio. The amplitudes are more likely to be close to zero in ordinary speech signals, and, therefore, uniform quantization brings larger relative error in case of low amplitudes value. Therefore, smaller quantization interval is considered at lower amplitudes. However, non-uniform quantizers are difficult to make and expensive. Therefore, an alternative approach is to first do the compression of the amplitude before passing it through uniform quantizer and consequent expansion at the receiver. This process is called companding technique.

3.3 Pulse Code Modulation (PCM)

Among pulse-modulated systems (PAM, PWM, PPM, and PCM), only PCM is of practical importance. The other systems are rarely used in practice. For this reason, only PCM is discussed in this chapter. Pulse code modulation (PCM) is basically analog-to-digital conversion of a special type where the information contained in the instantaneous samples of an analog signal is represented by digital words in a serial bit stream.

The three operations involved in PCM are: sampling, quantizing and encoding.

Sampling must be done at a rate of at least twice the highest frequency being transmitted. Sampling is a snapshot of the analog waveform at the instant of time.

Quantizing is rounding each PAM pulse value against a reference level.

Encoding is a process of approximating the amplitude sample by assigning to it the code value or labeling each PAM sample with the binary code for that level.

3.3.1 Sampling

The first step in the development of a PCM signal is sampling, which follows Nyquist sampling theorem. The sampling theorem states that:

If a band-limited signal is sampled at regular intervals of time and a rate equal to or higher than twice the maximum frequency of the signal, then the samples retain all the information of the original signal. The original signal can then be reconstructed by the use of a low-pass filter.

The sampling operation generates a flat-top PAM signal (Fig. 3.1). Sampling at a rate of f_s Hz can be accomplished by multiplying $x(t)$ by an impulse train $\delta_{T_s}(t)$, consisting of unit impulses repeating periodically every T_s seconds, where $T_s = 1/f_s$. This results in the instantaneous or flat-top sampled signal. The signal consists of samples spaced every T_s seconds. Here, it is noted that the input analog signal is a bandlimited signal which is sampled at higher rate so that the aliasing noise on the recovered analog signal is negligible.

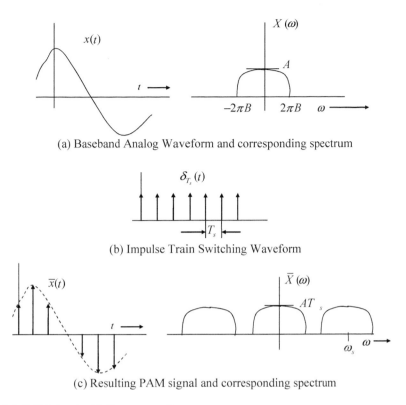

(a) Baseband Analog Waveform and corresponding spectrum

(b) Impulse Train Switching Waveform

(c) Resulting PAM signal and corresponding spectrum

Fig. 3.1 PAM signal with flat-top sampling

3.3.2   *Quantizing*

For quantizing, the amplitude of the message signal is limited to the range $(-V, V)$ is divided into L uniformly spaced intervals, each of the interval size, $\Delta v = \frac{2V}{L}$. A sample value is approximated by the midpoint of the interval in which it lies. The quantized samples are coded and transmitted as binary pulses. At the receiver, some pulses will be detected incorrectly. Hence, there are two sources of error in this scheme: quantization error and pulse detection error or bit error. In almost all practical schemes, channel coding can be used to correct some of the bit errors and consequently, reduce P_e. Thus it can be ignored.

The quantization operation is demonstrated in Fig. 3.2 for the case of M = 8 levels. This quantizer is called uniform quantizer (Fig. 3.3) as all the steps are of equal size. However, speech analog signals are more likely to have near zero amplitude values than at the extreme peak values. For example, in digitizing speech signals, the peak value may be 1 V, whereas weak signals may have voltage levels on the order of 0.1 V. For signals with such distribution, non-uniform quantizer is suitable for use. The goal of quantizer is to assign a sequence of bits to each of the

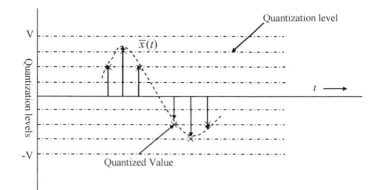

Fig. 3.2 Quantization process

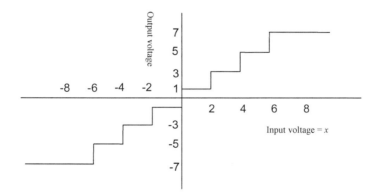

Fig. 3.3 Uniform quantizer characteristics

sample value. For the quantization process, we want to present to the coder a discrete voltage value. Suppose, our quantization step size is 0.1 V and our voltage measure for one sample was 0.37 V. This value needs to be rounded off to the nearest tenth which is 0.4 V, the nearest discrete value. It is noted here that there is a 0.03 V error, the difference between 0.37 and 0.4 V. This is said quantization error or quantization noise.

3.3.2.1 Signal to Quantization Noise Ratio, $(SNR)_Q$

The process of quantization can be interpreted as an additive noise process.

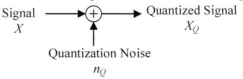

The **signal to quantization noise ratio** $(SNR)_Q = S/N$ is given as:

$$(SNR)_Q = \frac{Average\ Signal\ Power\ \{X\}}{Average\ Noise\ Power\ due\ to\ quantization\ \{n_Q\}}$$

Due to the following two main effects, the noise or distortion is produced in the PCM output:

- The first one is quantizing noise that is caused by the M-step quantizer at the PCM transmitter.
- The second effect is bit errors in the recovered PCM signal, caused by channel noise and improper filtering.

If the input analog signal is band limited and sampled at a higher rate so that the aliasing noise on the recovered signal is negligible, the ratio of the recovered analog peak signal power to the total average noise power is:

$$\left(\frac{S}{N}\right)_{pk\ out} = \frac{3M^2}{1 + 4(M^2 - 1)P_e} \tag{3.1}$$

where, $M = 2^n$, is the number of quantized levels used in the PCM system, P_e is the probability of bit error in the recovered binary PCM signal at the receiver DAC before it is converted back into an analog signal.

The ratio of the average signal power to the average noise power is

$$\left(\frac{S}{N}\right)_{out} = \frac{M^2}{1 + 4(M^2 - 1)P_e} \tag{3.2}$$

If we consider quantization noise only, then $P_e \approx 0$ (negligible), the above equations will become,

$$\left(\frac{S}{N}\right)_{pk\ out} = 3M^2 \text{ and } \left(\frac{S}{N}\right)_{out} = M^2$$

Putting the value of $M = 2^n$, we get $\left(\frac{S}{N}\right)_{Q,pk\ out} = 3.2^{2n}$ and $\left(\frac{S}{N}\right)_{Q,out} = 2^{2n}$

The **signal to quantization noise ratio** $(SNR)_Q = S/N$ in dB,

For peak $\left(\frac{S}{N}\right)_{Q,dB}$:

$$\left(\frac{S}{N}\right)_{Q,dBpk\ out} = 10\log\left(3.2^{2n}\right)$$

$$= 10\log 3 + 10\log 2^{2n}$$

$$= 4.77 + 6.02n \text{ dB}$$

$$\left(\frac{S}{N}\right)_{Q,dBpk\ out} = 4.77 + 6.02n \text{ dB} \tag{3.3}$$

For avg $\left(\frac{S}{N}\right)_{Q,dB}$:

$$\left(\frac{S}{N}\right)_{Q,dBpk\ out} = 10 \log \left(2^{2n}\right)$$
$$= 10 \log 2^{2n}$$
$$= 6.02n \text{ dB}$$

$$\left(\frac{S}{N}\right)_{Q,dBpk\ out} = 6.02n \text{ dB} \tag{3.4}$$

We can rewrite the above two equations in general form as

$$\left(\frac{S}{N}\right)_{dB} = 6.02n + \alpha, \tag{3.5}$$

where, n is the number of bits in the PCM word, $\alpha = 4.77$ for peak SNR.

Problem 3.1
In a communication-quality audio system, an analog voice-frequency (VF) signal with a bandwidth of 3200 Hz is converted into PCM signal by sampling at 7000 samples/s and by using a uniform quantizer with 64 steps. The PCM binary data are transmitted over a noisy channel to a receiver that has a bit error rate (BER) of 10^{-4}.

a. What is the null bandwidth of the PCM signal if a polar line code is used?
b. What is the average SNR of the recovered analog signal at the receiving end?

Solution
Given, the sampling frequency, $f_s = 7000$ samples/s, the number of quantization steps, $M = 64 = 2^n$, bit error rate (BER), $P_e = 10^{-4}$

a. The null bandwidth of PCM signal is

$$B_{null} = nf_s = 6(7000) = 42 \text{ kHz}$$

Note: If $sin\,x/x$ pulse shapes were used, the bandwidth would be

$$B_{null} = \frac{1}{2}nf_s = 21 \text{ kHz}$$

b. The average SNR of the recovered analog signal can be calculated as

$$\left(\frac{S}{N}\right) = \frac{M^2}{1 + 4(M^2 - 1)P_e} = \frac{64^2}{1 + 1.64} = 1552 = 31.9\,dB$$

3.3.3 Nonuniform Quantizing: Companding

The nonuniform quantizer is difficult to make, therefore expensive. Instead of using nonuniform quantizer, we will use companding technique. It is noted here that the companding and coding are carried out together. The effect of nonuniform quantizing is realized by first compressing signal samples using a compression amplifier and then using a uniform quantizer. If compression is done at the transmitter, expansion must be done using expander at the receiver output to restore their original values. The expandor characteristic is the inverse of the compression characteristic, and the combination of compressor and an expandor is called a compandor. The compression and later expansion functions are logarithmic. The logarithmic curve follows on of the two laws, the A-law and the μ-law. The equation for A-law can be expressed as:

$$y(x) = \left[\frac{A|x|}{1 + \ln(A)}\right] \qquad 0 \le |x| \le \frac{1}{A} \qquad\qquad (3.6)$$

$$y(x) = \left[\frac{1 + \ln A|x|}{1 + \ln A}\right] \qquad \frac{1}{A} \le |x| \le 1 \qquad\qquad (3.7)$$

where A = 87.6. The A-law is used with the E1 system. The μ-law can be expressed as:

$$y(x) = \frac{\ln(1 + \mu|x|)}{\ln(1 + \mu)} \qquad\qquad (3.8)$$

3.3.4 Encoding

The multiplexed PAM output is applied at the input of the encoder, which quantizes and encodes each sample into a group of n binary digits. A different type of encoder is available of which the digit-at-a-time encoder makes n sequential comparisons to generate an n-bit code word. The sample is compared with a voltage obtained by a combination of reference voltage proportional to 2^7, 2^6, 2^5, ..., 2^0. The reference voltages are conveniently generated by a bank of resistors R, 2R, 2^2R, ..., 2^7R.

3.4   PCM Transmission Bandwidth and the Output SNR

For binary PCM, the minimum number of samples required for a band limited signal of B Hertz is 2B (no aliasing) and each quantized sample is encoded into n bits. That means, a total of $n2B$ bits are required for PCM transmission. If a single bit is transmitted through a unit bandwidth (1 Hz), then the total transmission bandwidth of binary PCM is

$$B_{PCM} = n2B = nf_s \qquad (3.9)$$

The above equation shows that the transmission bandwidth for binary PCM depends on the bit rate and waveform pulse shape used to represent the bit. The minimum bandwidth depends on the pulse shape and can be obtained using $(\sin x)/x$ type pulse shape in generating the PCM waveform. However, due to using a more rectangular type of pulse, the bandwidth of the binary-encoded PCM waveform is larger than this minimum.

Once again, the output SNR follows the 6-dB law which is

$$\left(\frac{S}{N}\right)_{dB} = 6.02n + \alpha$$

where,

$$\alpha = 4.77 - 20\log(V/x_{rms}) \quad \text{(Uniform quantizing)}$$

Or, for sufficiently large input levels,

$$\alpha \approx 4.77 - 20\log[\ln(1+\mu)] \quad \text{(}\mu\text{-law companding)}$$

Or

$$\alpha \approx 4.77 - 20\log[1 + \ln A] \quad \text{(A-law companding)}$$

and n is the number of bits used in PCM word. Also, V is the peak value of the quantizer, and x_{rms} is the rms value of the input analog signal. It is noted that the output SNR of a uniform quantizer is a function of the input level whereas relatively insensitive to the input level for μ-law and A-law companding which is shown in Fig. 3.4. The ratio V/x_{rms} is called loading factor.

Problem 3.2
A compact disc (CD) records audio signals digitally by using PCM. Assume the audio signal bandwidth to be 15 kHz.

a. What is the Nyquist rate?
b. If the Nyquist samples are quantized into L = 65, 536 levels and then binary coded, determine the number of binary digits required to encode a sample.

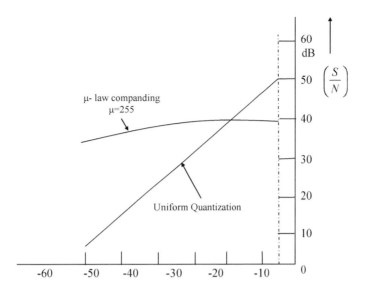

Fig. 3.4 Output $\left(\frac{S}{N}\right)_{out}$ of PCM system with or without companding technique

c. Determine the number of binary digits per second (bits/s) required to encode the audio signal.
d. For practical reasons discussed in the text, signals are sampled at a rate well above the Nyquist rate. Practical CDs use 44,100 samples per second. If L = 65,536, determine the number of bits per second required to encode the signal, the minimum bandwidth required to transmit the encoded signal.

Solution

a. The bandwidth is 15 kHz. The Nyquist rate is 30 kHz.
b. The quantization levels, $L = 65,536 = 2^n = 2^{16}$, so that 16 binary digits are needed to encode each sample.
c. The data rate, $R = nf_s = 16 \times 30 \times 10^3 = 480,000$ bits/s.
d. Practical rate, $44,100 \times 16 = 705,600$ bits/s.

3.5   Time Division Multiplexing of Digitized Speech Channels in PSTN

For the transmission over a common high speed medium, a number of signals (different bit streams) from different sources need to be multiplexed into a single higher rate stream. This is done using transmission time sharing by this number of signals by interleaving the pulse trains of various PCM signals in a specific order.

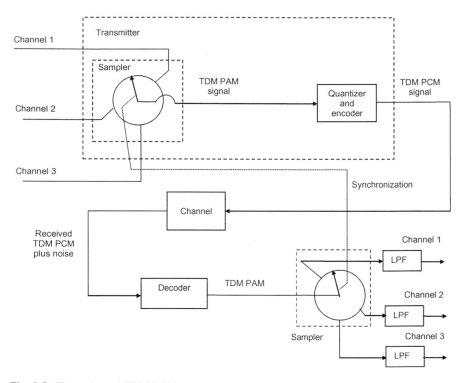

Fig. 3.5 Three-channel TDM PCM system

Fig. 3.6 T1 TDM format for one frame

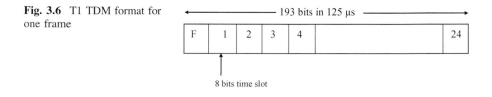

This process is known as time division multiplexing (TDM). After PCM, 24 different digitized voice channels are multiplexed together for voice transmission that results in a frame structure consisting of 193 bits in each 125 μs time interval that is 1.544 Mbps. This is the digital signal level 1 (DS1) signal that is carried on the T1 transmission lines (Figs. 3.5 and 3.6).

3.5.1   Higher Order Multiplexing

The DS1 (T1) signal with a data rate of 1.544 Mbps can be multiplexed with other DS1 signals to have higher order carrier. For example, the DS2 signal is obtained

Fig. 3.7 T1 system signaling format

through multiplexing of 4 DS1 lines. The output rate is 6.312 Mbps which is a slight higher than the expected due to overhead (framing bits) and also to equalize timing variations. Since the 4 streams might be generated at different clocks, the output stream is to be formed at a slight higher rate to keep away from dropping bits. The output stream can, therefore, contain extra "stuff" bits in designated positions, in the bit stream to boost up the faster rate for the DS2 signal. This process is called plesiochronous (almost synchronous) multiplexing. The problem with this scheme is that it is difficult at any point in its transmission path to access any particular voice channel without passing through the entire demultiplexing process. The synchronous digital hierarchy (SDH) is a newer scheme that uses synchronous multiplexing stages, using a master clock signal.

A note on the DS1 signal: The DS1 structure allows data to be transmitted at 1.544 Mbps [24 channels $\times$ 8 + 1 bit for framing = 193 bits/frame, 8000 frames/ s, therefore, data rate = 193 $\times$ 8000 = 1.544 Mbps]. Since all eight bits are used for transmission instead of seven bits used in the earlier version, the signaling channel provided by the eight bit is no longer available. Every sixth frame, therefore, has 7 $\times$ 24 = 168 information bits, 24 signaling bits and 1 framing bit. In all the remaining frames, there are 192 information bits and 1 framing bit. In view of the "robbed-bits" in the 6th and 12th frames of each super frame, the data rate is dropped to 56 Kbps. Notice that for voice, it is barely perceptible if losing the least significant bit one out of every 6 samples. Finally, note that it is possible to lease a dedicated T1 line for data transmission at a higher rate (Fig. 3.7).

3.6 Differential Pulse Code Modulation (DPCM)

Generally, it is observed that successive samples are close to the same value when audio or video signals are sampled. This means that there is a lot of redundancy in the signal samples and consequently, the bandwidth of a PCM system is wasted due to the transmission of redundant sample values. One way to reduce this redundant transmission and thus decrease the bandwidth requirement is to transmit PCM signals corresponding to the difference in adjacent sample value which is called differential pulse code modulation (DPCM). At the receiver, the present sample value is regenerated by using the past value plus the update differential value that is received over the differential system. Moreover, the present value can be estimated from the past values by using a prediction filter.

Figure 3.8 uses the predictor to get a differential pulse amplitude-modulated (DPAM) signal which is quantized and encoded to generate the DPCM signal.

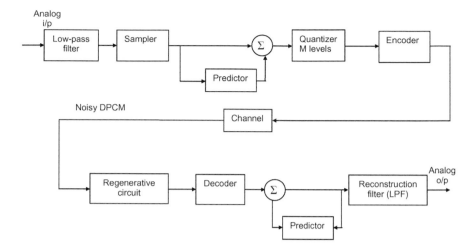

Fig. 3.8 DPCM, using prediction from samples of input signal

The recovered analog signal at the receiver output will be the same as that at the system input, plus accumulated quantizing noise.

It can be shown that DPCM, like PCM, follows the 6-dB rule

$$\left(\frac{S}{N}\right)_{dB} = 6.02n + \alpha$$

where, α is in between $-3 \prec \alpha \prec 15$ for DPCM speech and n is the number of quantizing bits. Unlike companded PCM, the α for DPCM varies over a wide range, depending on the properties of the input analog signal.

3.7 Delta Modulation

The Delta modulation (DM) is the simplest form of DPCM and the circuit for DM is shown in Fig. 3.9. The functions of the subtractor and two-level quantizer are implemented by using a comparator so that the output is $\pm V_c$ (binary). In this case, the DM signal is a polar signal. In the receiver, the DM signal may be converted back to an analog signal approximation to the analog signal at the system input. This is accomplished by using an integrator for the receiver that produces a smoothed waveform corresponding to a smoothed version of the accumulator output waveform that is present in the transmitter. The integrator itself constitutes a low pass filter.

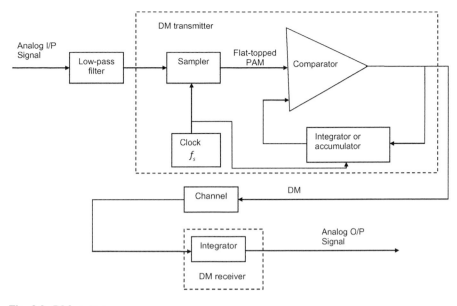

Fig. 3.9 DM system

3.8   **Integrated Services Digital Network (ISDN)**

The integration of voice, video, telex, fax and data traffic in a single digital network found major market application. The Integrated Services Digital Network (ISDN) is an example of this integration. ISDN offers circuit switched connections and packet switched connections, in increments of 64 Kbps. ISDN in its basic form allows 128 Kbps service over a pair of standard telephone copper wires. The 144 Kbps payload is broken down into two 64 Kbps bearer channels and one 16 Kbps control signaling (call set-up, etc.) channel. Thus, the subscriber can use basic access channel with a 192 Kbps data including synchronization and framing bits.

In ISDN terminology, a physical connection is made between a DTE and the ISDN network termination equipment (DCE). The physical connection specifies an 8-pin RJ45 connector to provide data transmission in each direction between the terminal devices and the network devices. The electrical signaling is in the form of baseband pulses. With an ISDN connection, the user can send data and/or voice over the network to another ISDN terminal. More generally, ISDN allows circuit-switched service, packet-switched service, or dedicated connection. Each 64 Kbps channel can support data, digitized voice, or a combination of lower data rate and digitized voice intended for the same endpoint.

In ISDN, voice and data are both treated as digital circuit-switched services. The voice is digitized at the terminal to be integrated with the data.

3.9 Sample Questions

1. What does the sampling theorem tell us concerning the rate of sampling required for an analog signal?
2. Consider an audio signal with spectral components in the range 300–3000 Hz. Assume that a sampling rate of 70,000 samples per second will be used to generate a PCM signal.

   a. For SNR = 30 dB, what is the number of uniform quantization levels needed?
   b. What data rate is required?

3. Why should PCM be preferable to DM for encoding analog signals that represent digital data?
4. For a PCM signal, determine L if the compression parameter $\mu = 100$ and the minimum SNR required is 45 dB. Determine the output SNR with this value of L. Remember that L must be a power of 2, that is, $L = 2n$ for a binary PCM.
5. A television signal (video and audio) has a bandwidth of 4.5 MHz. This signal is sampled, quantized, and binary coded to obtain a PCM signal.
6. Determine the sampling rate if the signal is to be sampled at a rate 20% above the Nyquist rate.
7. If the samples are quantized into 1024 levels, determine the number of binary pulses required to encode each sample.
8. Determine the binary pulse rate (bits per second) of the binary-coded signal, and the minimum bandwidth required to transmit this signal.

References

1. L.W. Couch, *Digital & Analog Communication Systems*, 8th edn. ISBN: 978-9332518582
2. B.P Lathi, *Modern Digital and Analog Communication Systems*, 3rd edn. ISBN: 978-0195110098

Chapter 4
AM, Angle Modulation and Digital Modulation Systems

4.1 Introduction

Modulation is the process of encoding information on to the carrier in a manner suitable for transmission. It generally causes a shift in the range of frequencies in the signal. In carrier modulation, a baseband message signal is translated to a bandpass signal at frequencies that are very high when compared to the baseband frequency. In this modulation, one of the basic parameters amplitude, phase, or frequency of a high frequency carrier is varied in accordance to the message signal. Demodulation is the process of extracting the baseband message from the modulated signals corrupted by noise so that it may be processed and interpreted by the intended receiver. This chapter describes various modulation and demodulation techniques that are used in communication systems.

4.2 Amplitude Modulation (AM)

Amplitude modulation is a modulation scheme in which the amplitude of a carrier sinusoidal waveform $c(t)$ is changed in accordance to the modulating signal $m(t)$. The equation for AM wave can be written as (Fig. 4.1)

$$S_{AM}(t) = [A + m(t)] \cos \omega_c t \qquad (4.1)$$

4.2.1 AM Spectrum

The complex envelop of an AM signal is $g(t) = [A + m(t)]$, and the corresponding spectrum of this envelop is $G(\omega) = A\delta(\omega) + M(\omega)$. The spectrum of $S_{AM}(t)$ is expressed as

© The Author(s) 2018   43
M. A. Matin, *Communication Systems for Electrical Engineers*,
SpringerBriefs in Electrical and Computer Engineering,
https://doi.org/10.1007/978-3-319-70129-5_4

Fig. 4.1 Amplitude
modulated (AM) wave

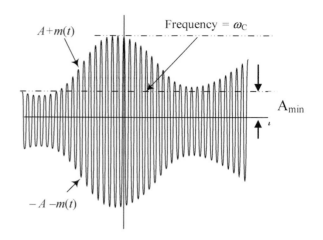

$$S(\omega) = \frac{1}{2}[M(\omega + \omega_c) + M(\omega - \omega_c)] + \pi A[\delta(\omega + \omega_c) + \delta(\omega - \omega_c)]$$

where, $m(t)$ is real, $M^*(\omega) = M(-\omega)$ and $\delta(\omega) = \delta(-\omega)$ were used.

Consider the baseband (message) signal is a single sinusoid of frequency ω_m. Then the modulated signal will consist of two sinusoids: the component of frequency $\omega_c + \omega_m$ (USB) and the component of frequency $\omega_c - \omega_m$ (LSB). If the message signal is a band of frequencies, suppose the magnitude spectrum of the modulating signal happens to be a triangular function, then the spectrum will be as follows which is shown in Fig. 4.2.

Problem 4.1 An AM wave is represented by the expression

$$v = 3(1 + 0.8\,Cos\,6120\,t)Sin\,211 \times 10^4 t \, V$$

(a) What are the minimum and maximum amplitudes of the AM wave?
(b) What frequency components are contained in the modulated wave and what is the amplitude of each components?

Solution

The AM wave equation is given as:

$$v = 3(1 + 0.8\,Cos\,6120\,t)Cos\,211 \times 10^4 t \, V$$

The standard AM wave equation is: $S_{AM}(t) = [A + m(t)]\cos\,\omega_c t$, Comparing the given equation with the standard one, we get, $A = 3$,

$$m(t) = 2.4\,Cos\,6120\,t, \omega_c = 211 \times 10^4, f_c = 211 \times 10^4/2\pi = 336\,KHz$$

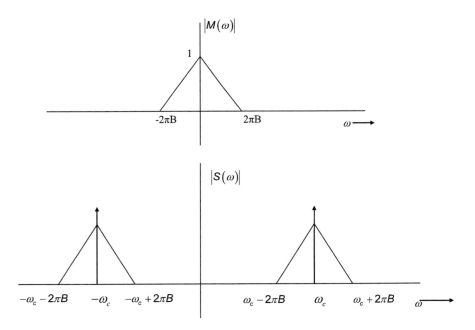

Fig. 4.2 Spectrum of AM signal

The carrier amplitude, $A = 3$ V, the modulating signal amplitude, $A_m = 2.4$ V modulation depth $= A_m/A = 2.4/3 = 0.8$, Carrier frequency, $f_c = 336$ KHz, modulating signal frequency, $f_m = 6120/2\pi = 974$ Hz.

Minimum amplitude of AM wave, $V_{min} = A - A_m = 0.6$ V

Maximum amplitude of AM wave, $V_{max} = A + A_m = 5.4$ V

The AM wave will contain three frequencies viz

$f_c - f_m$	f_c	$f_c + f_m$
335,026 Hz	336,000 Hz	336,974 Hz

The amplitudes of the three components of AM wave are:

$\frac{mA}{2}$	A	$\frac{mA}{2}$
1.2 V	3 V	1.2 V

4.2.2   Double Sideband Suppressed Carrier (DSB-SC)

In DSB-SC, the amplitude of the modulated signal is varied in proportion to the message signal and the time-domain representation of the modulated signal can be written as Eq. (4.2). As the carrier frequency component does not exist in the

Fig. 4.3 DSB-SC
modulation

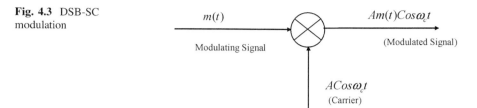

modulated signal, this type of AM is called double sideband, suppressed carrier
modulation (Fig. 4.3).

$$S_{DSB-SC}(t) = m(t) \cdot c(t)$$

$$S_{DSB-SC}(t) = Am(t) \cos \omega_c t \tag{4.2}$$

where, $c(t) = A \cos(2\pi f_c t)$. Note that, $S_{DSB-SC}(t)$ is the modulated signal that
results from a cosine carrier wave $c(t)$ being multiplied by a modulating baseband
signal, $m(t)$ which contains the information to be transmitted. The spectrum of
DSB-SC wave can be expressed as:

$$S(f) = \frac{A}{2}[M(f - f_c) + M(f + f_c)]$$

If the message signal is a band of frequencies, suppose the magnitude spectrum
of the modulating signal is a triangular function, then the spectrum of the double
sideband suppressed carrier (DSB-SC) can be illustrated as in Fig. 4.4.

Special Case: consider modulating signal is a single frequency sinusoid

Consider the modulating signal is $m(t) = A_m \cos \omega_m t$, where, A_m is the peak
amplitude of modulating signal, ω_m is the modulating signal frequency, unit is in
radians/second. The modulating signal $m(t)$ contains all the information that has to be
sent. By inserting this expression into Eq. (4.2), the equation for $S_{DSB-SC}(t)$ becomes,

$$S_{DSB-SC}(t) = AA_m \cos(\omega_m t) \cos(\omega_c t) \tag{4.3}$$

Equation (4.3) can be re-written as

$$S_{DSB-SC}(t) = \frac{A.A_m}{2}[\cos(\omega_c + \omega_m)t + \cos(\omega_c - \omega_m)] \tag{4.4}$$

It is noted from the above Eq. 4.4 that the modulated signal has two components
of which each one is shifted in frequency by the modulating frequency and there is
no carrier component. This type of modulation is therefore called double sideband,
suppressed carrier amplitude modulation (DSB-SC AM). The two sidebands are
called the upper sideband (USB), and the lower sideband (LSB).

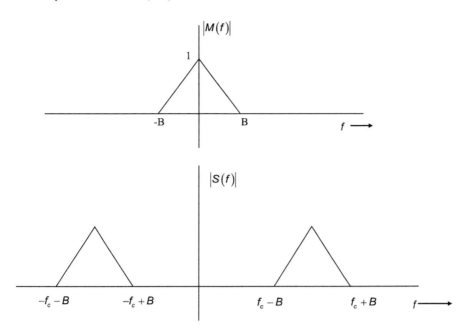

Fig. 4.4 Spectrum of DSB-SC signal

Consider, the amplitude of the carrier is 1, then the $S_{DSB-SC}(t)$ will become

$$S_{DSB-SC}(t) = \frac{A_m}{2}[\cos(\omega_c + \omega_m)t + \cos(\omega_c - \omega_m)] \qquad (4.5)$$

The Spectra of the double sideband suppressed carrier (DSB-SC) is given by

$$S_{DSB-SC}(t) \Leftrightarrow \pi\frac{A_m}{2}\left[\begin{array}{l}\{\delta(\omega + (\omega_c + \omega_m)) + \delta(\omega - (\omega_c + \omega_m))\} + \\ \{\delta(\omega + (\omega_c - \omega_m)) + \delta(\omega - (\omega_c - \omega_m))\}\end{array}\right] \qquad (4.6)$$

The carrier has higher frequency than the modulating signal, typically in the order of at least 100. For example, the radio station broadcasting AM signals where the carrier frequency is about 1000 kHz and the modulating signal is about 5 kHz. Figure 4.5 shows all the frequencies that could exist in the modulated signal. In DSB-SC, there is no carrier present at ω_c.

Figure 4.5 shows two sidebands in the DSB-SC spectrum and each of these has exactly the same information of the original modulating signal. A scheme in which only one sideband is transmitted is known as single-sideband (SSB) transmission, which requires only one half of the DSB signal. Additionally, the sidebands are always symmetric about the carrier frequency. Due to having the same peak voltage in the two sidebands given by $\pi A_m/2$, the average power is same. This power can be calculated using the following equation, $\frac{V_{rms}^2}{R}$.

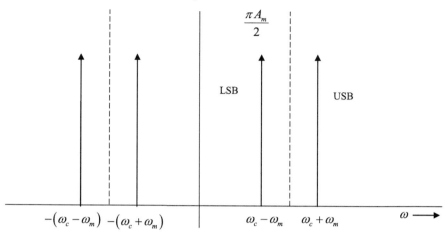

Fig. 4.5 Spectrum of DSB-SC AM wave for single frequency modulating signal

4.2.3   AM Modulation Index

The m is a constant defined as the ratio of the modulating signal amplitude to the un-modulated carrier amplitude:

$$m = \frac{A_m}{A} \tag{4.7}$$

It can be represented in terms of percentage modulation obtained by multiplying m by 100% to obtain the number in percentage form:

$$\%modulation = m \times 100\%$$

The modulation index m should be in between 0 and 1. If m = 0, the resultant waveform is just the carrier wave and the amplitude of carrier wave remains unchanged, which is A. The carrier wave is not modulated. If m > 1, the resultant waveform is over modulated and is distorted.

Problem 4.2 The maximum and minimum voltages of an AM wave are V_{max} and V_{min} respectively, show that the modulation factor m can ne represented as

$$m = \frac{V_{max} - V_{min}}{V_{max} + V_{min}}$$

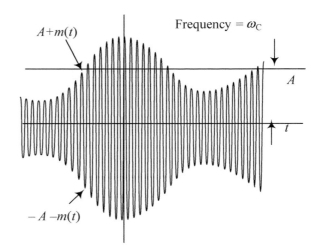

Fig. 4.6 AM modulated wave

Solution

Let the amplitude of the carrier wave is A and the modulating signal amplitude A_m. Then the modulation index, $m = \frac{A_m}{A}$. It is clear from the Fig. 4.6 that $V_{\text{max}} = A + A_m$ [also as, $S_{AM}(t) = (A + A_m \cos \omega_m t) \cos \omega_c t$]

$$V_{\text{min}} = A - A_m$$

Now, solving the above equations, we can get, $A = \frac{V_{\text{max}} + V_{\text{min}}}{2}$ and

$$A_m = \frac{V_{\text{max}} - V_{\text{min}}}{2}$$

The modulation index, $m = \frac{A_m}{A} = \frac{V_{\text{max}} - V_{\text{min}}}{V_{\text{max}} + V_{\text{min}}}$ (Proved).

4.2.4 Power of an AM Signal

There are three components of an AM waveform: the upper and lower sidebands and the carrier frequency. The sideband power is the useful power and the carrier power is the wasted power. To get the total power, all these three need to be added together. The total power in the modulated signal is

$$P_{AM} = P_c + P_{USB} + P_{LSB} \tag{4.8}$$

To compute the power of amplitude modulated signal, root mean square (rms) values of all three components must be used.

$$P_{AM} = \frac{A^2}{2R} + \frac{A_m^2}{8R} + \frac{A_m^2}{8R} = \frac{A^2}{2R} + \frac{A_m^2}{4R}, \quad P_{AM} = \frac{A^2}{2R}\left[1 + \frac{1}{2}\frac{A_m^2}{A^2}\right]$$

$$P_{AM} = \frac{A^2}{2R}\left(1 + \frac{m^2}{2}\right) = P_c\left(1 + \frac{m^2}{2}\right) \tag{4.9}$$

$$P_{AM} = P_c(1 + \tfrac{m^2}{2})$$

In case of zero percentage modulation ($m = 0$), the modulated signal has the power of only unmodulated carrier (as not modulated), which results in

$$P_{AM} = P_c = \frac{A^2}{2R}\left(1 + \frac{0}{2}\right) = \frac{A^2}{2R} \tag{4.10}$$

Problem 4.3 A carrier wave of 500 watts is subjected to 100% amplitude modulation. Determine (a) power of modulated wave (b) power in sidebands.

Solution

Given, carrier power = 500 W, depth of modulation = 100% = 1
   The power in AM waveform is,

$$P_{AM} = P_c(1 + \frac{m^2}{2})$$

Power of modulated wave, $P_{AM} = 500(1 + \frac{1}{2}) = 500 \times \frac{3}{2}$ W = 750 W
Sideband power, $P_{SB} = Power\ of\ AM\ wave\ -\ carrier\ power = (750 - 500)$ W
Sideband power = 250 W.

Problem 4.4 The antenna current of an AM transmitter is 8 A when only carrier is sent but it increases to 8.93 A when the carrier is modulated due to superimposing the message signal. Find the percentage modulation.

Solution

The power in AM waveform is,

$$\frac{P_{AM}}{P_c} = (1 + \frac{m^2}{2}), \quad P_{AM} = P_c\left(1 + \frac{m^2}{2}\right)$$

$$\frac{P_{AM}}{P_c} = \frac{I_{AM}^2 R}{I_c^2 R} = \left(1 + \frac{m^2}{2}\right)$$

$$\frac{8.93^2}{8^2} = \left(1 + \frac{m^2}{2}\right)$$

$$\frac{m^2}{2} = 0.246, \quad m = 0.701 = 70.1\%$$

4.2.5 AM Signals Generation

AM signals are basically analogous to double sideband suppressed carrier (DSB-SC) modulation. In AM, a constant is added first with the modulating signal and then modulated by the carrier. Using DSB-SC modulation technique, the AM signals can be generated. Several methods have been devised for AM signal generation. One method is to multiply message signal by the carrier waveform whose output is proportional to the product of two inputs. The other method is to use nonlinear devices such as semiconductor diode or transistor. The multiplication operation required for modulation can be replaced by switching operation. Figure 4.7 shows a switching modulator, where the switching action is provided by a diode.

The message and carrier signal are amalgamated to generate modulated waveform using switching modulator circuit. The combined sources drive the diode circuit whose output is the half–wave rectified signal. The spectrum of this half–wave rectified signal contains different harmonics which are multiples of ω_c.

$$v_R = [\cos \omega_c t + m(t)] \left[\frac{1}{2} + \frac{2}{\pi} \left(\cos \omega_c t - \frac{1}{3}\cos 3\omega_c t + \frac{1}{5}\cos 5\omega_c t - \ldots \right) \right]$$

$$v_R = \frac{1}{2}\cos \omega_c t + \frac{2}{\pi}m(t) \cos \omega_c t + \ldots \qquad (4.11)$$

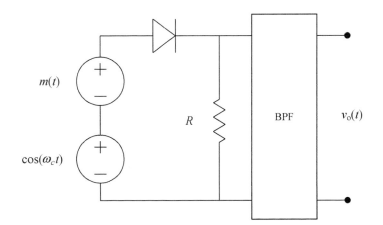

Fig. 4.7 Switching modulator

The bandpass filter tuned to ω_c suppresses all other terms, yielding the desired AM signal at the output.

4.2.6   Demodulation Methods of AM

Demodulation is identical to modulation except the output filter in the demodulator circuit. In the modulator, the multiplier output is passed through a bandpass filter tuned to ω_c, whereas in the demodulator, the multiplier output is passed through a low-pass filter. The AM signal can be demodulated coherently by a locally generated carrier in case of coherent demodulation or detection. For coherent demodulation, the receiver must generate a carrier in phase and frequency synchronism with the incoming carrier. However, coherent or synchronous demodulation of AM can be used for any value of modulation index but expensive, hence will rarely use in practice. Here, two noncoherent methods of AM are explained: (1) rectifier detection, and (2) envelop detection.

4.2.6.1   Rectifier Detector

The rectifier detector circuit is shown in Fig. 4.8 where the AM signal is applied to the diode. Due to the diode characteristics (Fig. 4.8), the negative part of the AM wave will not pass through the circuit. The output across the resistor is a half-wave rectified version of the AM signal which contains various components at different frequencies around 0, ω_c, $2\omega_c$, $3\omega_c$, ... etc., including the message, $m(t)$ and a dc component, A. Note that all components except the message and dc components are filtered by the low pass filter (LPF). The capacitor at the end of the circuit is to block dc component so that the output of the signal is a scaled version of $m(t)$ without dc component.

4.2.6.2   Envelope Detector

In an envelope detector, the output takes the envelope of the modulated signal. The circuit shown in Fig. 4.9 is an envelope detector which is the modification of the rectifier detector. The diode is either forward–biased (when the AM signal is higher with respect to the capacitor voltage), or reverse–biased (when the AM signal is below the capacitor voltage). During forward bias, the diode acts like a short circuit (in ideal case) and the capacitor is charging. The voltage across the capacitor follows the voltage of the source (while source voltage increases, capacitor voltage also increases) until the AM signal value is higher than the voltage developed

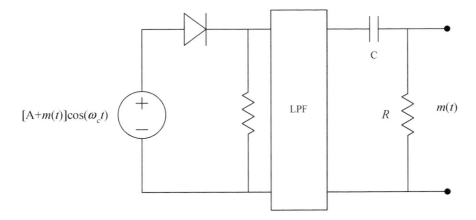

Fig. 4.8 Rectifier detector for AM

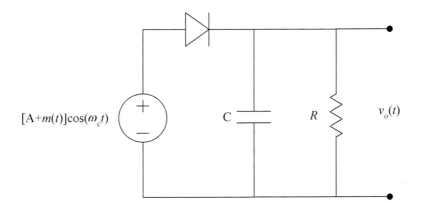

Fig. 4.9 Envelop detector for AM

across the capacitor. In case of reverse–bias, the diode acts like an open circuit and the capacitor voltage is reduced by being discharged through the resistor. If the value of the time–constant of the capacitor and resistor $\tau = RC$ is suitable (not too large or too small), the charging and discharging of the capacitor results in a signal that follows the message signal with some small ripples. The main advantage of this form of AM demodulator is that it is very simple and cheap.

The Fig. 4.10 illustrates both AM modulated signal and demodulated signal in time–domain.

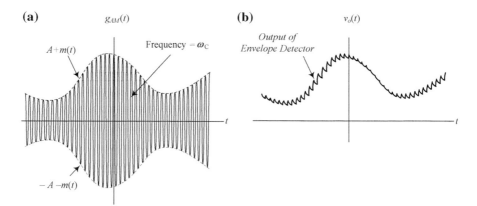

Fig. 4.10 a AM modulated signal and **b** Demodulation of AM signals

4.3   Angle Modulation (FM and PM)

Angle modulation schemes include both frequency modulation (FM) and phase modulation (PM). Both FM and PM belong to the class of non-linear modulation schemes. This family of modulation schemes is featured by their high-bandwidth requirements and good performance in the presence of noise.

4.3.1   Frequency Modulation (FM)

In FM, the message signal $m(t)$ controls the frequency f_c of the carrier (not the phase as in PM). Consider the carrier, $v_c = A\,Cos(\omega_c t + \Phi_c)$ where, $(\omega_c t + \Phi_c)$ represents the angle of the carrier. Then for FM, we can write:

$$v_{FM}(t) = A \cdot \cos\left[\omega_c t + k_f \int_{-\infty}^{t} m(\alpha)d\alpha\right],$$

where, k_f is a parameter that specifies how much change in the frequency occurs for every unit change of $m(t)$.

$$v_{FM}(t) = A\,\cos(2\pi(f_c + frequency\,deviation)t) \qquad (4.12)$$

Note that the frequency deviation will depend on $m(t)$.

Now, we can write the above equation as, $v_{FM} = A\,Cos(2\pi f_i t) = A\,Cos(\varphi_i)$, where, φ_i is the instantaneous angle and $\varphi_i = 2\pi f_i t$, then $\frac{d\varphi_i}{dt} = 2\pi f_i$ or $f_i = \frac{1}{2\pi} \cdot \frac{d\varphi_i}{dt}$ i.e. frequency is proportional to the rate of change of angle.

If f_c is the unmodulated carrier and f_m is the modulating frequency, then we can deduce that $f_i = f_c + \Delta f_c \cos(\omega_m t) = \frac{1}{2\pi} \cdot \frac{d\varphi_i}{dt}$, Δf_c is the peak deviation of the carrier. Hence, we have $\frac{1}{2\pi} \cdot \frac{d\varphi_i}{dt} = f_c + \Delta f_c \cos(\omega_m t)$

$$\text{i.e. } \frac{d\varphi_i}{dt} = 2\pi f_c + 2\pi \Delta f_c \cos(\omega_m t) \tag{4.13}$$

Integrating both sides of Eq. (4.13),

$$\varphi_i = \omega_c t + \frac{2\pi \Delta f_c \sin(\omega_m t)}{\omega_m}$$

We can rewrite the above equation as, $\varphi_i = \omega_c t + \frac{\Delta f_c}{f_m} \sin(\omega_m t)$.

Hence for the FM signal, $v_{FM} = A \cos\left(\omega_c t + \frac{\Delta f_c}{f_m} \sin(\omega_m t)\right)$.

The ratio $\frac{\Delta f_c}{f_m}$ is called the modulation index and can be represented as m_f. If $m(t)$ is considered as a 'single tone modulating signal' as of the following form, $m(t) = A_m \cos(\omega_m t)$, the equation for FM signal can be expressed as Bessel Functions and can be written as

$$v_{FM} = A \sum_{n=-\infty}^{n=+\infty} J_n(m_f) \cos(\omega_c + n\omega_m)t, \tag{4.14}$$

where, $J_n(m_f)$ are Bessel functions of the first kind. Expanding the equation, we get

$$v_{FM} = AJ_0(m_f)Cos(\omega_c t) + AJ_1(m_f)Cos(\omega_c + \omega_m)t + AJ_{-1}(m_f)Cos(\omega_c - \omega_m)t \\ + AJ_2(m_f)Cos(\omega_c + 2\omega_m)t + AJ_{-2}(m_f)Cos(\omega_c - 2\omega_m)t + \ldots$$

The amplitudes drawn in Fig. 4.11 are completely arbitrary since $J_n(m_f)$ is unknown.

4.3.2 FM Signal Spectrum and Bandwidth

From the FM equation, it is evident that the FM modulated signal contains a carrier component and an infinite number of sidebands of frequencies, $\omega_c \pm \omega_m$, $\omega_c \pm 2\omega_m$, ..., $\omega_c \pm n\omega_m$ (Fig. 4.11). The strength of the nth sideband at $\omega = \omega_c + n\omega_m$ is $J_n(m_f)$. For a sufficiently larger value of n, $J_n(m_f)$ is negligible, and there are only a finite number of significant sidebands. It can be seen that $J_n(m_f)$ is negligible for $n \succ m_f + 1$. Hence, the number of significant sidebands is $m_f + 1$. The bandwidth of the FM is given as

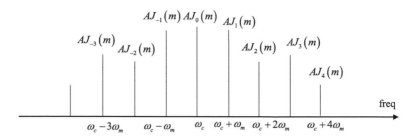

Fig. 4.11 FM signal Spectrum

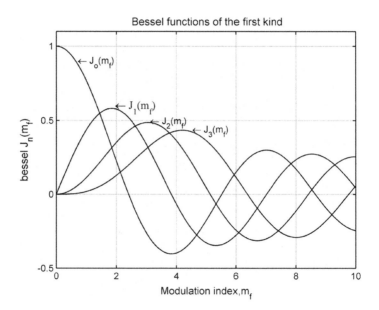

Fig. 4.12 Plot of the Bessel functions of the first kind, integer order

$$B_{FM} = 2nf_m = 2(m_f + 1)f_m$$
$$= 2\left(\frac{\Delta f_c}{f_m} + 1\right)f_m$$
$$= 2(\Delta f_c + f_m)$$

The information extracted from the Bessel function of first kind (Fig. 4.12) put into tabular form for integer or fractional values of m_f which is shown in Table 4.1. From the table of Bessel functions, it can be observed that for small value of m_f, i.e. if $(m_f \leq 0.3)$ there is only the carrier and two significant sidebands, i.e. $BW = 2f_m$. FM with $m_f \leq 0.3$ is referred to as narrowband FM (NBFM). For $m_f \succ 0.3$, there

Table 4.1 Bessel-function of the first kind

m_f	Bessel-function order, n									
	J_0	J_1	J_2	J_3	J_4	J_5	J_6	J_7	J_8	J_9
0.00	1.00									
0.25	0.98	0.12								
0.5	0.94	0.24	0.03							
1.0	0.77	0.44	0.11	0.02						
1.5	0.51	0.56	0.23	0.06	0.01					
2.0	0.22	0.58	0.35	0.13	0.03					
2.5	−0.5	0.5	0.45	0.22	0.07	0.02	0.01			
3.0	−0.26	0.34	0.49	0.31	0.13	0.04	0.01			
4.0	−0.40	−0.07	0.36	0.43	0.28	0.13	0.05	0.02		
5.0	−0.18	−0.33	0.05	0.36	0.39	0.26	0.13	0.05	0.02	
6.0	0.15	−0.28	−0.24	0.11	0.36	0.36	0.25	0.13	0.06	0.02

are more than 2 significant sidebands. As m_f increases the number of sidebands increases. This is referred to as wideband FM (WBFM).

Problem 4.5 A carrier is frequency modulated with a sinusoidal signal of 4 kHz resulting in a maximum frequency deviation of 5 kHz. Find

(i) modulation index
(ii) bandwidth of the modulated signal.

Solution

The following parameters are given:
Maximum frequency of the sinusoidal input signal is, $f_m = 4\,\text{KHz}$
Maximum frequency deviation, $\Delta f_c = 5\,\text{KHz}$
Modulation index $= \frac{\Delta f_c}{f_m} = \frac{5}{4}$
Bandwidth of the modulated signal $= 2(\Delta f_c + f_m) = 18\,\text{KHz}$.

4.3.3   Power in FM Signal

The equation for FM (Eq. 4.14) is, $v_{FM} = A \sum_{n=-\infty}^{n=+\infty} J_n(m_f) \cos(\omega_c + n\omega_m)t$.

It is seen in the above equation that the peak value of the components is $AJ_n(m_f)$ for the nth component. The signal normalised average power is $= \left(\frac{V_{pk}}{\sqrt{2}}\right)^2 = (V_{rms})^2$,

then the nth component power is $\left(\frac{AJ_n(m_f)}{\sqrt{2}}\right)^2 = \frac{A^2}{2} \cdot J_n^2(m_f)$.

As, $\frac{A^2}{2}$ is the FM transmitter output power, the nth component power is,

$$P_n = J_n^2(m_f) \times P_{trans}$$

Hence, the total power in the infinite spectrum is

$$\text{Total power}, P_{total} = \sum_{n=-\infty}^{n=+\infty} \frac{\left(AJ_n(m_f)\right)^2}{2}$$

Rearranging the above equation, we get $P_{total} = \sum_{n=-\infty}^{n=+\infty} J_n^2(m_f) \times P_{trans}$.

4.3.4   FM Signal Generation

There are two basic methods for generating FM signals- direct method and indirect method. Direct FM method uses an active device in implementing a voltage to frequency (V/F) conversion function. One such device is varactor diode. The characteristic of varactor diode is equivalent to a capacitor in reverse biased condition. Furthermore, the equivalent capacitance of the varactor diode varies based on the reverse bias voltage applied to it. Used in conjunction with a tuned circuit, the varactor diode can convert an input signal voltage into a varying oscillator output frequency.

Consider now, an analogue message input, $m(t) = V_m \cos(\omega_m t)$.

In the direct method, the instantaneous carrier frequency is directly varied in accordance with the input modulating signal. As the input $m(t)$ varies from $+V_m \rightarrow 0 \rightarrow -V_m$, the output frequency will vary from a maximum, through f_c, to a minimum frequency (Fig. 4.13). The f_{out} can be written as $f_{out} = f_c + \alpha m(t)$, i.e. the deviation depends on $m(t)$. Considering that the maximum and minimum input amplitudes are $+V_m$ and $-V_m$ respectively, then

$$f_{max} = f_c + \alpha V_m$$
$$f_{min} = f_c - \alpha V_m$$

Fig. 4.13 FM generation

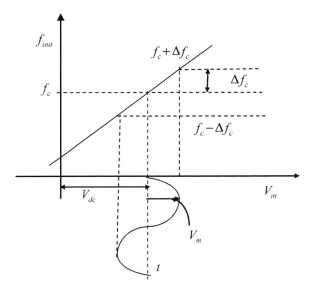

The peak-to-peak deviation is $f_{max} - f_{min}$, but more importantly for FM, the peak deviation Δf_c is,

$$\Delta f_c = \alpha V_m$$

Hence, Modulation Index, $\beta = \frac{\Delta f_c}{f_m} = \frac{\alpha V_m}{f_m}$

4.3.5   FM Demodulation-General Principles

The information in an FM signal resides in the instantaneous frequency. Hence, a demodulator is required which can translate the instantaneous frequency into message signal. Therefore, an FM demodulator or frequency discriminator is essentially a frequency-to-voltage converter (F/V) which generates an output voltage, V_{out} proportional to the instantaneous frequency at the input, f_{in}.

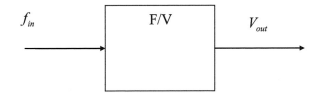

If the input is FM, the output will be $m(t)$, the analogue message signal. One can use an operational amplifier differentiator as an . A simple tuned circuit followed by an envelop detector can also serve as a frequency detector. Because of low cost and superior performance especially in case of low SNR, FM demodulator using PLL is widely used today.

4.3.5.1 Tuned Circuit

The tuned circuit has a frequency to amplitude transfer characteristic over the bandwidth. It is tuned so the f_c, the nominal input frequency is on the slope, not at the centre of the tuned circuits. As the frequency of the FM signal varies about f_c on the tuned circuit slope, it changes its position on the slope of the tuned circuit. The amplitude of the output changes in proportion to the deviation from f_c. Thus, the FM signal is successfully converted to AM. This is then converted into message signal by the diode detector circuit (Fig. 4.14).

4.3.5.2 Phase-Locked Loop (PLL)

The cost of PLL FM demodulator is low. It provides superior performance, especially when the SNR is low. Therefore, it is widely used (Fig. 4.15).

The operation of the PLL is similar to that of a feedback system. In a typical feedback system, the signal fed back tends to chase the input signal and to minimize the difference in phase between the feedback signal and the input FM wave. If the signal fed back is not equal to the input signal, an error occur (the difference known as error signal) which will change the signal fed back until it is close to the input signal. A PLL operates on a similar principle, except that the quantity fed back and compared is not the amplitude, but the phase. It tracks the incoming signal angle and instantaneous frequency. The VCO adjusts its own frequency until it is equal to that of the input FM

Fig. 4.14 FM demodulator
(tuned circuit)

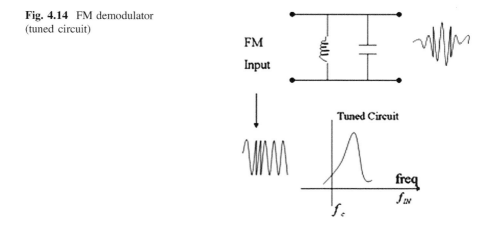

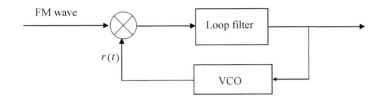

FM wave

r(t)

Loop filter

VCO

Fig. 4.15 FM demodulator (PLL circuit)

signal frequency. At this point, the frequency and phase of the two signals ate in synchronized (except for a possible difference of a constant phase).

4.3.6 Frequency Modulation Versus Amplitude Modulation

In FM, the amplitude of the carrier signal is kept constant (constant envelope signal), whereas the frequency of carrier is changed based on the amplitude of the modulating message signal. Since message signal is translated to frequency variations rather than amplitude variations in FM signals, it is less vulnerable to atmospheric disturbance that causes rapid fluctuations in the amplitude of the received radio signal. Moreover, FM offers superior qualitative performance in fading in compare to AM but requires more complicated transmitter and receiver.

4.3.7 Phase Modulation (PM)

In Phase Modulation (PM), the phase of the carrier signal is modified based on the message signal. The phase modulated signal can be expressed as

$$g_{PM}(t) = A \cdot \cos\left[\omega_c t + k_p m(t)\right],$$

where A is a constant, ω_c is the carrier frequency, $m(t)$ is the message signal, and k_p is a parameter that specifies the amount of change in the angle per unit change of $m(t)$. The phase of this PM modulated signal is

$$\theta_{PM}(t) = \omega_c t + k_p m(t),$$

and the corresponding frequency is,

$$\omega_i(t) = \omega_c + k_p \frac{dm(t)}{dt} = \omega_c + k_p \dot{m}(t).$$

It is seen form the above equation that there is a linear relationship between the frequency of a PM signal and the derivative or the slope of the message signal.

4.3.8 Relation Between PM and FM

FM and PM are interrelated. From the phase and frequency relationship in PM and FM, it is observed that if $m(t)$ in the PM signal is replaced with $\int_{-\infty}^{t} m(\alpha)d\alpha$, it will give an FM signal. On the other hand, if $m(t)$ in the FM signal is replaced with $\frac{dm(t)}{dt}$, PM signal will be generated.

4.4 Generalized Receiver: Superheterodyne Receiver

The function of the receiver is to extract the baseband modulating signal from the received signal that might be suffered great attenuation due to the presence of obstacle and noise while transmission. Usually, we are expecting the exact replica of the modulating signal at the receiver output that was present at the transmitter input. The receivers can be classified into two: the tuned radio-frequency (TRF) receiver and the superheterodyne receiver.

There is a number of cascaded high-gain RF bandpass stages in the TRF receiver that are simultaneously tuned to the received frequency, followed by an appropriate detector circuit (an envelop detector, a product detector, an FM detector, etc.). The TRF is the simplest type receiver but not very popular, because of difficulties in designing tunable RF stages which can select the desired station as well as reject the adjacent channel stations. Moreover, gain is not uniform over a wide range of radio frequencies and to have sufficiently small stray coupling between the output and input of the RF amplifying chain so that the chain will not become an oscillator at f_c. TRF receivers are often used to measure time-dispersive (multipath) characteristics of radio channels.

The superheterodyne receiving technique is used in most of the receivers. The technique is accomplished with a mixer where the selected radio frequency is mixed with the high frequency. In this process, beats are produced and the mixer generates a frequency, called the intermediate frequency (IF) band and then extracting the baseband modulating signal by using the appropriate detector. The production of fixed IF is the salient feature of the superheterodyne receiver. The superheterodyne receiver is used for all types of bandpass signal reception, such as television, FM, AM, satellite, cellular, and radar signals. The RF amplifier tunes the desired signal and amplifies it to swamp out noise in the mixer stage. The RF filter provides a desired bandpass characteristic that reject adjacent channel signals and noise, but the main adjacent channel rejection is done by the IF filter (Fig. 4.16)

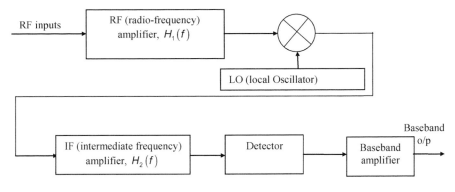

Fig. 4.16 Superheterodyne receiver

4.5 Digital Modulation Techniques

Digital communication is popular as it requires cost effective digital circuitry. The flexibility of digital approach arises because of merging digital data from digital sources with digitized data derived from different analog sources and transmits together over a high speed communication link. It uses only a finite number of symbols for communication, the minimum number being two (the binary case). The original information could be represented as digital bit stream of 0's and 1's. These bits are translated into symbols using a bit-to-symbol mapping, which in this case could be as simple as mapping the bit 0 to the symbol +1, and the bit 1 to the symbol −1. The digital baseband modulation methods are also known as line coding which is to transfer a digital bit stream over a baseband channel using a pulse train, i.e. a discrete number of signal levels, by directly modulating the voltage or current on the channel. The most common examples are unipolar, non-return-to-zero (NRZ) and manchester coding. Among other desirable properties, a line code should have the following properties:

1. Small transmission bandwidth
2. Power efficiency
3. Error detection and correction capability
4. Favourable power spectral density
5. Adequate timing content
6. Transparency.

4.5.1 Binary Modulated Bandpass Signaling

Thus far, we have discussed baseband digital systems, where signals are transmitted directly without shift in the frequencies of the signal. It is also realized that

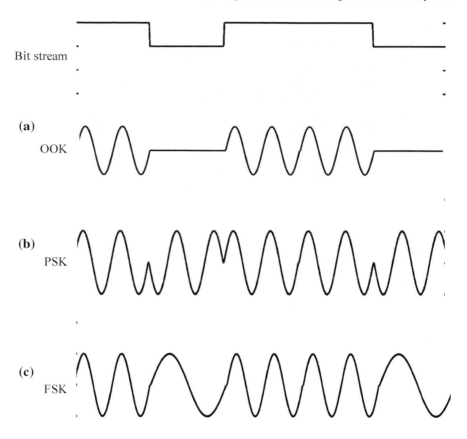

Bit stream

(a)

OOK

(b)

PSK

(c)

FSK

Fig. 4.17 **a** OOK, **b** PSK and **c** FSK signal

bandpass digital communication signals are produced by using baseband digital signals to modulate a carrier. Two basic forms of modulation exist: amplitude modulation and angle modulation. In amplitude modulation, the carrier amplitude is varied in proportion to the modulating signal (i.e., the baseband signal) whereas in angle modulation, the carrier frequency or phase of the carrier is varied according to the modulating signal. The amplitude modulation scheme of transmitting binary data is known as on-off keying (OOK) or amplitude shift keying (ASK) (Fig. 4.17).

The ASK/OOK signal can be described as

$$s(t) = m(t)\cos(\omega_c t) = \begin{cases} A\,\cos(\omega_c t) & m(nT_b) = A(\text{``1''}) \\ 0 & b(nT_b) = 0(\text{``0''}) \end{cases}$$

The FSK signal

$$s(t) = \begin{cases} A\cos(\omega_1 t) & m(nT_b) = A \\ A\cos(\omega_0 t) & m(nT_b) = -A \end{cases}$$

Bandwidth of BPSK Signal

The BPSK signal is described by

$$s(t) = m(t)\cos(\omega_c t) = \begin{cases} A\cos(\omega_c t) & m(nT_b) = A(\text{"1"}) \\ A\cos(\omega_c t + \pi) & m(nT_b) = -A(\text{"0"}) \end{cases}$$

To evaluate the spectrum of $s(t)$, let us consider the worst case which requires the widest bandwidth. The worst scenario happens when the digital modulating waveform has frequent transitions. In this case $m(t)$ would be a square wave, as shown in Fig. 4.18. Here, the binary data is represented by +1 V and −1 V for binary 1 and 0 respectively and the signalling rate is $R = \frac{1}{T_b}$ bits/s. The power spectrum of the square wave modulating signal can be evaluated by using Fourier series analysis (discussed in Chap. 1).

$$P_m(f) = \sum_{n=-\infty}^{n=\infty} |c_n|^2 \delta(f - nf_0) = \sum_{n=-\infty}^{n=\infty} \left[\frac{\sin(n\pi/2)}{n\pi/2}\right]^2 \delta\left(f - n\frac{R}{2}\right) \qquad (4.15)$$

where, $f_0 = 1/(2T_b) = R/2$. The PSD of $s(t)$ can be expressed in terms of the PSD of $m(t)$ by evaluating the autocorrelation of $s(t)$-that is

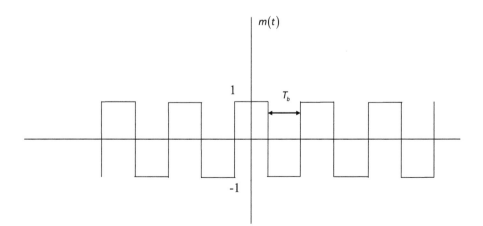

Fig. 4.18 Modulating signal, $m(t)$ is a square wave

$$R_s(\tau) = \frac{1}{2}R_m(\tau)\cos\,\omega_c\tau$$

The PSD is obtained by taking the Fourier transform of both sides of the above equation. Using the real signal frequency transform theorem, we get

$$P(f) = \frac{1}{4}[P_m(f - f_c) + P_m(f + f_c)] \qquad (4.16)$$

The value of Eq. 4.15 is plugged into Eq. 4.16, we obtain the PSD for the BPSK signal:

$$P(f) = \frac{1}{4}\sum_{n=-\infty}^{n=\infty}\left[\frac{\sin(n\pi/2)}{n\pi/2}\right]^2\left\{\delta\left(f - f_c - n\frac{R}{2}\right) + \delta\left(f + f_c - n\frac{R}{2}\right)\right\} \qquad (4.17)$$

The spectral shape that comes from the worst-case deterministic modulation is essentially the same as that obtained when random data are used; however, for the random case, the spectrum is continuous. The PSD result for polar NRZ signal is:

$$P_{polarNRZ}(f) = A^2 T_b\left(\frac{\sin\,\pi f T_b}{\pi f T_b}\right)^2$$

The corresponding PSD of BPSK signal is

$$P(f) = \frac{1}{4}T_b\left[\frac{Sin\pi T_b(f - f_c)}{\pi T_b(f - f_c)}\right]^2 + \frac{1}{4}T_b\left[\frac{Sin\pi T_b(f + f_c)}{\pi T_b(f + f_c)}\right]^2 \qquad (4.18)$$

when the data rate is $R = 1/T_b$ bits/s.

4.5.2   Multilevel Modulated Bandpass Signaling

In multilevel signalling, digital inputs with more than two levels are allowed on the transmitter input. Figure 4.19 shows that the serial bit stream is transformed into

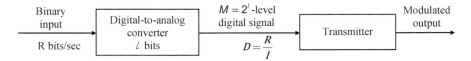

Fig. 4.19 Multilevel digital transmission system

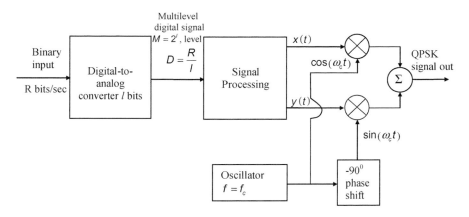

Fig. 4.20 Modulator for Generalized Signal Constellation

multilevel signals using digital to analog converter (DAC). If the transmitter is a PM transmitter with an input of $M = 4$ level digital signal, the M-ary phase shift keying is generated at the transmitter output.

A 2-BIT DIGITAL TO ANALOG CONVERTER	
Binary Input ($1 = 2$ bits)	Output Level (V)
11	+3
10	+1
00	−1
01	−3

M-ary PSK (MPSK) can also be generated by using two quadrature carriers modulated by the x and y components of the complex envelop; in that case,

$$g(t) = A_c e^{j\theta(t)} = x(t) + jy(t)$$

Where the permitted values of x and y are

$$x_i = A_c \cos \theta_i \text{ and } y_i = A_c \sin \theta_i$$

For the permitted phase angles θ_i, $i = 1, 2, \ldots, M$, of the MPSK signal. The situation is illustrated in Fig. 4.20, where the signal processing circuit implements the above equations. Figure 4.21 gives the relationship between the permitted phase angles θ_i and the (x_i, y_i) components for QPSK and $\frac{\pi}{4}$ QPSK signal constellations.

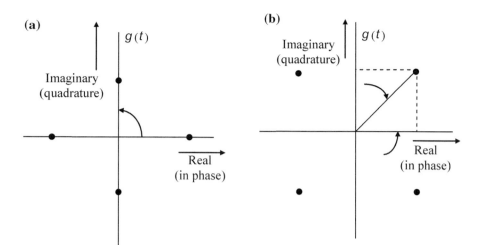

Fig. 4.21 QPSK and π/4 QPSK signal constellations (permitted values of the complex envelop)

4.6 Sample Questions

1. Explain amplitude modulation (AM). Derive the voltage equation of an AM wave.
2. An AM voltage signal with a carrier frequency of 1150 kHz has a complex envelop $g(t) = A_c[1 + m(t)]$, $A_c = 500\,\text{V}$, and the modulation is a 1-KHz sinusoidal test tone described by $m(t) = 0.8\sin(2\pi1000t)$. Evaluate the voltage spectrum for this AM signal.
3. Draw the block diagram of superhetrodyne AM receiver and briefly explain each block.
4. Draw and explain the diode detector circuit used for AM.
5. Derive the expression for the total transmitted power in AM and analyze the expression for power savings in SSB transmission.
6. The maximum peak-to-peak voltage of an AM wave is 16 mV while the minimum peak-to-peak voltage is 8 mV. Find the percentage modulation.
7. Define the term modulation factor and explain the importance of modulation factor in communication system.
8. For a baseband signal $m(t) = \cos 1000t$, find the DSB-SC signal, and sketch its spectrum.
9. In the early days of radio, AM signals were demodulated by a crystal detector followed by a low-pass filter and a dc blocker, as shown in the following figure. Assume a crystal detector to be basically a squaring device. Determine the signals at points a, b, c and d. Point out the distortion term in the output $y(t)$. Show that if $A \gg |m(t)|$, the distortion is small.

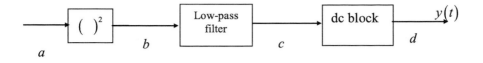

10. Define frequency modulation (FM). Derive an expression for FM Signal and analyze the expression to calculate the bandwidth. Explain the method of generating an FM wave.

11. An angle-modulated signal with carrier frequency $\omega_c = 2\pi \times 10^5$ is described by the equation

$$v_{FM} = 10Cos(\omega_c t + 5Sin\,300t + 10Sin\,2000\pi t)$$

(a) Find the power of the modulated signal.
(b) Find the frequency deviation.
(c) Estimate the bandwidth of FM modulated signal.

12. What are the differences among angle modulation, PM and FM?
13. How are binary values represented in amplitude shift keying, frequency shift keying and phase shift keying?
14. What is QPSK? Draw the constellation diagram of QPSK.

Chapter 5
Telecommunication Systems

5.1 Introduction

Telecommunication can be defined as exchange information using electrical signals or electromagnetic waves among station equipments that are far apart from each other. Any station equipment located in one place of the world, can communicate with any other station equipment in other place of the world. The station equipment could be a computer, a FAX machine, a teleprinter, a data terminal, and so on. Billions of station equipments, all over the world are engaged in processing and transferring the information as a telephone conversation or a file exchange between two computers or a message between two terminals. The rapid growth of subscribers and genuine need for new services, telecommunication becomes a part of our life. In reality, telecommunication turns into the alternative means of the expensive physical transportation. Therefore, telecommunication sector has an explosive growth in last decade and will be undoubtedly continuing to do so.

In the old days, the telecommunication links and switches were mostly designed for voice communication. However, modern telecommunication needs new facilities with large bandwidth switched data networks and large communication satellites with small and cheap ground station. This chapter describes the fundamentals of telecommunication network which provides a good understanding of today's telecommunication systems.

5.2 Evolution of Modern Telecommunications

The telegraphic signals transmission over wires was the first scientific achievement in the arena of modern telecommunication. In 1837, Samuel F.B. Morse's developed a version of the electrical telegraph which was demonstrated in 1844, over 40 miles between Baltimore and Washington. In 1845, Morse founded a telegraph

© The Author(s) 2018   71
M. A. Matin, *Communication Systems for Electrical Engineers*,
SpringerBriefs in Electrical and Computer Engineering,
https://doi.org/10.1007/978-3-319-70129-5_5

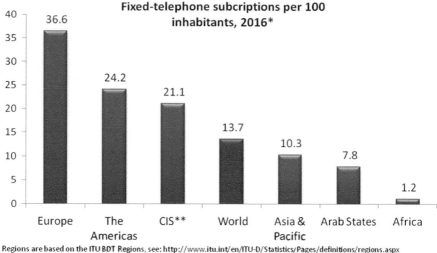

Regions are based on the ITU BDT Regions, see: http://www.itu.int/en/ITU-D/Statistics/Pages/definitions/regions.aspx
Note: * Estimate ** Commonwealth of Independent States
Source: ITU World Telecommunication /ICT Indicators database

Fig. 5.1 The world fixed-telephone subscriptions statistics

company based on his technology. In 1849, the first slow telegraph printer link was established. In 1874, Ban dot discovered a "Multiplexes" system which allows up to six signals from telegraph machines to be combined together for transmission over the same line. The use of telegraphy did greatly enhance the dissemination of news and personal messages between towns. In 1876, Alexander Graham Bell invented a telephone system and in 1877 established the Bell telephone company. Since then, telephony has become flourished. The automatic switching was another important advance in the development of telephony. In 1889, Almon B. Strowger invented the first automatic step-by-step circuit switching system and applied for US patent which was issued in 1891. The first two crossbar switches went into services in 1938 in NewYork city. With this technological progress, infrastructure deployment and falling price have brought unexpected growth in telephony sector. Figure 5.1 shows the statistics for the fixed-telephone subscriptions across the globe.

5.3   Simple Telephone Communication

In a telephone communication, information is exchanged between calling and called subscribers. Telephone set comprises a microphone associated with the speaker, is able of both receiving and sending but not in simultaneous manner. If calling subscriber is transmitting at one instant of time, the called subscriber is receiving and vice versa. Figure 5.2 gives a simple telephone circuit where the microphone

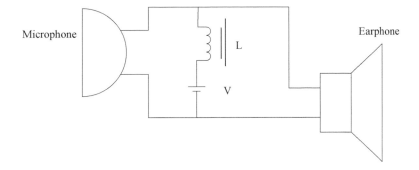

Fig. 5.2 A simplex telephone circuit

and the speaker are the transducer elements. Microphone converts audio signal into electrical signals whereas speaker converts electrical signals into audio signals.

The electrical power is received from the local exchange via two copper wires for the operation of ordinary home telephone. This subscriber line, that carries speech signals is a twisted pair called a local loop.

5.3.1   Subscriber Loop Signaling

In a telephone network, subscriber loop signaling depends on the telephone set type that is being used. The subscriber loop signaling for a rotary dial telephone and dual tone multi-frequency are shown in Figs. 5.3 and 5.4 respectively.

In pulse dialing, a train of pulses is utilized to imply a digit of the subscriber number. The number of pulses in the train is equal to the digit value apart from "zero" digit which is represented by 10 pulses. Line is rapidly disconnected and reconnected in sequence with one pulse for digit value "1", two pulses for digit value "2", etc. Each pulse lasts 0.1 s. Inter-digit pause (IDP) must be >0.5 s otherwise current digit can combine with the previous digit. This dialing takes about 12 s to dial a 7-digit number. Pulse dialing is restricted to signaling between the exchange and the subscriber and no signaling is possible end-to-end, i.e.

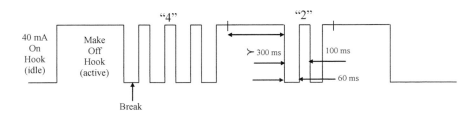

Fig. 5.3 Dial pulse address signaling

Fig. 5.4 DTMF signaling

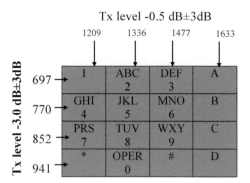

between two subscribers. The rotary dial is replaced by a push button keyboard. 'Touching' a button generates a 'tone' which is a combination of two frequencies, one from the lower band and the other from the upper band. For example, pressing the push button 9 transmits 852–1477 Hz. It is faster than the pulse dialing and reduces call setup time and chances of error or interference.

5.3.2 Signaling Tones to the Exchange from the Telephone

There is a need of some signalling tones by which a telephone conversation can be started, maintained and released. In addition to these functions, a subscriber can know what happening to his or her attempted calls is, and also knows if there is network congestion or the other party is busy. There are five types of signalling tones which work at automatic exchanges. These are:

1. Dial Tone
2. Ringing Tone
3. Busy Tone
4. Number unobtainable Tone
5. Call-in-progress Tone.

Dial Tone—This tone is being heard if someone lifts the handset indicating the readiness of the exchange to accept the dialed numbers from the calling subscriber. This tone is sent to the calling subscriber by exchange. If dial tone is not received by calling party by any reason then the call may not be completed by exchange. So, the subscriber should start dialing digits after hearing a dial tone. The dial tone is 33 or 50 or 400 Hz continuous tone. Usually the 400 Hz signal is modulated with a 25 or 50 Hz signal.

Ringing Tone—If the called subscriber is available after dialing a number, the exchange will send a ringing tone to the called party. Simultaneously, this tone is received by the calling party. The ringing tone is nothing but the familiar double-ring pattern. The two rings in the double-ring pattern are alienated by a time-gap of 0.2 s and two double-ring patterns by a gap of 2 s. The ring burst has time duration of 0.4 s. The frequency of the ringing tone is 133 or 400 Hz.

Busy Tone—If the network is busy, the switching instrument or line i.e. resources are not available or called party is attending another call then a busy tone is sent to calling the party. The busy tone is a bursty 400 Hz signal with equal 0.375 s on/off periods.

Number unobtainable Tone—This is a continuous 400 Hz signal. This is sent to the calling subscriber when called party is out of service or disconnected, and due to an error in dialing.

Call-in-progress Tone—This is a 400 or 800 Hz having 0.1 s on/off periods. This indicates that the dialed number that means call is in progress.

5.3.3   Telephone Numbering Plan

Following the international numbering plan, subscribers from different countries can call each other. The numbering format for international telephone number is shown in Fig. 5.5. The trunk code and the subscriber number together create a unique identification for dialling subscriber at the national level.

International Prefix: An international prefix or international access number is used for international calls. It tells the network that the connection is to be routed via an international telephone exchange to another country. The international prefix '00' should be dialed first followed by the telephone number.

Country code: The country code contains one to four numbers that define the country of subscriber. Country codes are not needed for national calls because their purpose is to make the subscriber identification unique in the world.

Trunk/area code: The trunk code defines the area inside the country where the call is to be routed. To make a call from one region to other region, the area code needs to be dialed as well as the trunk prefix.

Subscriber number: The subscriber number in a fixed telephone network is a unique identification of the subscriber inside a geographical area. To connect to a certain subscriber, the same number is dialed anywhere in the area. This is the combination of exchange code and the line number.

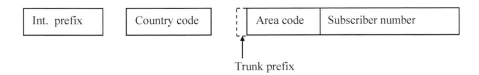

Fig. 5.5 International telephone number format

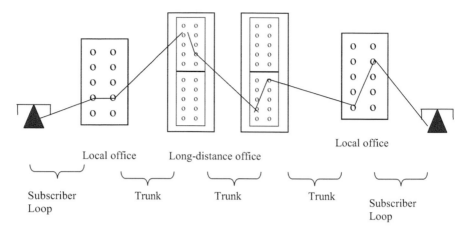

Fig. 5.6 Basic telephone call

5.4 Telephone Call and Charging Plan

5.4.1 Telephone Call Set-up

To establish a call, the following sequences of event are involved in telecommunication system (Fig. 5.6).

1. Calling customer lifted the handset from the cradle; phone goes to off hook that closes the circuit to the local office ("looping the circuit").
2. Local office detects the "loop" and indicates readiness with dial tone.
3. Once calling customer hears the dial tone, he dials the number of called party.
4. The network converts ("translates") the phone number to a physical equipment address.
5. The network checks on the called party status and decides on a routing for the connection.
6. If the connection is possible, the called party is alerted. Large 20 Hz alternating current is applied to the line ("ringing current").
7. "Ring tone" is returned to the caller.
8. The called party picks up the handset and closes his/her loop.
9. Exchange detects second loop and "trips" or stops ringing, then establishes call and voice conversion starts.
10. Any party can terminate the call by hanging up, and exchange clears connection.

5.4.2   Charging Plan

The subscribers in telecommunication share a component of capital cost as well as individual call cost which includes the operating costs in establishing and maintaining the calls. The capital cost includes line plant, switching systems, building and land. Considering all these factors, a charging plan for a telecommunication service levies three different charges on a subscriber:

An initial charge for providing a network connection.

- A rental or leasing charge
- Charges for individual calls made.

The charge for individual telephone call is time based and distance related. The rate for a local call is different from a national call, and international calls. It is simple and straightforward in Europe and the countries following European telecommunication practice. A meter with a stepping motor is connected to every subscriber line at the local exchange. The number of pulses per second activating the meter is originated from the exchange code or the area code of the dialed number. For example: the local call under the same exchange generates 1 pulse (one step) per minute, a call to nearby city 3 pulses, or a call carried over to distant city 10 pulses per minute. In case of international calls, it checks the digits of the country code and then set the meter to generate the pulses at an even higher rate. All completed calls must be charged; hence the metering circuitry must also sense call supervision to act in response to call completion—that is, "off hook" to "on hook" condition. The meter pulses starts at "off hook" condition and stops "on hook" condition i.e. termination of the call. This is referred to as pulse metering or "flat rate" billing. Such billing information is recorded by verifying first the calling subscriber number and then the called number. Also, of course, there is a necessity of timing for call duration. In detailed billing, individual subscriber-dialed calls can be determined whenever the bulk local charge is separated.

5.5   Telecommunication Network

The modern telecommunication systems can be best described in terms of a network which includes the basic elements, network infrastructure and the control systems.

5.5.1   Elements of Telecommunication Network

The users in telecommunication network are called subscribers (who pay the subscription charge). The user information can be of any form such as voice or data,

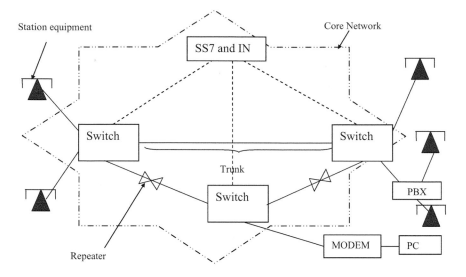

Fig. 5.7 Telecommunication network architecture

and subscribers can use different access point which connects the population of users to a serving network node, and a core portion which interconnects the set of serving network nodes. A public switched telecommunication network (Fig. 5.7) can be described using the following architectural components:

1. Subscribers
2. Switching system
3. Trunks
4. Subscriber lines
5. Station equipment
6. Transmission equipment
7. Network signaling and control.

   Subscriber: Customers who disburse the subscription charge is called subscribers. The devices that attach to the end office in the network are the subscriber's device. The subscriber's devices are the transceiver that is responsible for both sending and receiving information. Most subscriber devices to PSTN network are telephones.

   Switching System: The function of telecommunication switching system is to transmit message from one terminal device to another terminal device (Fig. 5.8). A switching system is a set of switching elements, which is also referred to as

Fig. 5.8 Model of a switching network

switching centers. The switching centers in the network receive the messages and control and forwards to the required destination. A switching center of the telephone network comprises a switching network and its control and support equipments. It is called a central office. The central office receives the control signals, messages or conversations. It is then forwarded to the required destination, after modification if required.

In computer communication, packet switching or message switch is used whereas in telephone network, circuit switching is used. Some practical switching systems are step-by-step, crossbar switch, digital switching systems, electronic switching system, etc.

Trunks: The link between switches is called trunks. Trunks carry multiple voice circuits using either FDM or synchronous TDM.

Subscriber loop: The line between the subscriber and the end office or the local office is termed as the subscriber loop or local loop. The local loop connections use twisted-pair wire. The subscriber line transmits the information and control signals among subscribers and switching centers.

Station Equipment: The station equipment or instrument is a transmitter or receiver that is liable for sending information or decoding or inverting received information to the original one. Station equipments in the telephone network have progressed from analog telephones to digital handsets and is connected to telephone lines. The other devices could also be connected to telephone lines, such as computer terminals which are used for data transmission.

5.5.2 Technologies in Telecommunication Network

The three technologies needed for communication through the network are (1) transmission system, (2) switching system, and (3) signaling system.

5.5.2.1 Transmission Systems

Transmission is the process of carrying information between end points of a system or a network. In telecommunication network, the station equipment or the instruments generate information signals for transmission to the destination. The transmission link carries the information and control signals between the station equipments and switching centers. In general, a communication link between two distinct locations is established through a number of transmission lines in tandem. The transmission line comprises of copper cable, transverse-screen cable, coaxial cable, optical fibre or in the case of a radio system, the air, over which the telecommunications takes place. The line can be characterized by its bandwidth, attenuation and the propagation delay. However, for long distance transmission, repeater and regenerator equipment are required along the transmission line to boosts and corrects the signal compensating for the loss and distortions incurred while propagating through the line.

5.5.2.2 Switching Systems

A switching system or an exchange is a set of switching elements that are organized and managed like so as to establish a communication link between inlet and outlet pairs. The early switching systems were manual and operator driven. To overcome the limitations of operator manned switching systems, automatic exchanges came into existence which is classified into two: electromechanical and electronic (Fig. 5.9). However, the electromechanical switching systems have been replaced by Electronic Switching System (ESS) or Electronic Automatic Exchange (EAX) which is also called Stored Program Control (SPC). The world's first electronic switching system, known as No. 1 ESS, was commissioned in 1965. Since then, there is a rapid and continuous growth of electronic switching system and stored program control in versatility and range of services. Today, SPC is a standard feature of all the electronic exchanges.

Modelling of Switching Systems

A telecommunication network carries traffic that is generated by a large number of subscribers connected to the network. Subscribers generated calls in a random fashion. The call generation and therefore the behavior of the network or the switching systems can be described as a random process. A random process or a stochastic process is one in which one or more parameters fluctuate with time in a

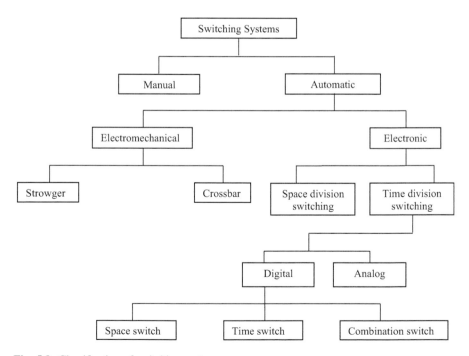

Fig. 5.9 Classification of switching systems

manner that the instantaneous values of the parameters are not measurable accurately but are estimable with certain probability. The statistical properties of a random process can be obtained either by observing its behavior over a very long period of time, or by observing simultaneously, a very large number of statistically identical random sources at any given instant of time. The statistical properties obtained using the first method are known as time statistical parameters, and the ones obtained using the second method are called ensemble statistical parameters. The word ensemble denotes the collection of sources. The behavior of a switching system can also be modeled as a random process or stochastic process. For example, the number of servers busy simultaneously is a discrete random variable. The time at which a server becomes busy or free also exhibits a random behavior, and hence the entire system can be modeled as a random process.

5.5.2.3 Signalling System

Signalling (Fig. 5.10) connects different switching systems, transmission systems and subscriber equipments in a telecommunication network to facilitate the network to operate en bloc.

The signalling system is an essential building block for automatic telephone services which exchanges signalling information efficiently between subscribers. Signalling offers the interface between different national systems. The introduction of signalling system was a significant movement in developing the PSTN. The functions of signalling are to establish, maintain and terminate a call. It also provides network management functions. In PSTN, in-band signaling, the signaling information pertaining to a particular speech channel is carried within the same channel that the telephone call itself is using. In-band signaling implies reduced bandwidth for speech transmission.

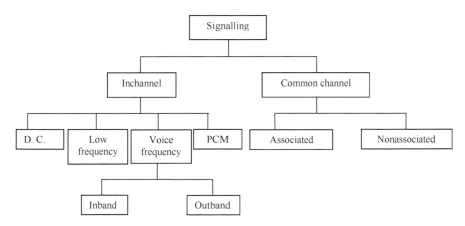

Fig. 5.10 Signalling techniques

There is a separate dedicated channel for signaling in Out-of-band signaling. Out-of-band signaling has been used since the introduction of Signaling System No. 6 (SS6) in 1970s. Then Signaling System No. 7 (SS7) is introduced in 1980 and it became the standard for signaling among exchanges ever since.

5.6   Telecommunication Design and Implementation

5.6.1   Telecommunication Access Network Architecture: Fiber to the x

Fiber to the x (FTTX): is a common term for broadband network design using optical fiber which offer all or part of the local loop used for last mile telecommunications. It comes in different flavor such as FTTC, FTTB, FTTH etc., depending on how close is the optical fiber terminated to the end user.

FTTH (Fiber-to-the-home): Fiber arrives at the periphery of the living space, such as a box on the outer wall of a home. Passive optical networks and point-to-point Ethernet are architectures that deliver triple-play services over FTTH networks directly from an operator's central office.

FTTB (Fiber-to-the-building, business, or basement): Fiber reaches the boundary of the building, such as the basement in a multi-dwelling unit, with the final connection to the individual living space being made via alternative means, similar to the curb or pole technologies.

FTTN (Fiber-to-the-node, neighborhood): Fiber is terminated in a street cabinet, possibly miles away from the customer premises, with the final connections being copper. FTTN is often an interim step toward full FTTH.

FTTC (fiber-to-the curb, closet or cabinet): This is similar to fiber-to-the-node (FTTN), but the street cabinet or pole is closer to the user's premises, typically within 1000 ft (300 m), within range of high bandwidth copper technologies such as wired Ethernet or IEEE 1901 power line networking and wireless Wi-Fi technology.

5.6.2   Metro (SONET/SDH, Metro Ethernet) and Core Network Architecture

The core networks and metropolitan area networks (also called metro networks) in today are almost exclusively depends on optical fiber systems. The exploitation of optical fiber systems in the access networks is done through fixed/wireless convergence. Optical-fiber-based networks are apparently the governing technology that to be considered as the backbone of the future Internet.

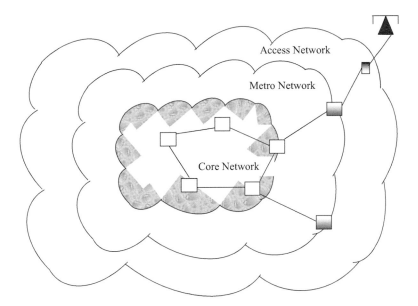

Fig. 5.11 Core and access network

Today, there is a substantial growth of telecommunication networks and network designers are always rethinking for a cost-effective, secure, and reliable telecommunications network. The research today is focused on how network dynamicity can be significantly enhanced, how the cost of ownership can be reduced, and how the industrial cost of network services can be decreased. Figure 5.11 shows the network architecture for metro/regional and core/backbone segments.

5.7   Telecommunication Traffic Engineering

Traffic engineering is the starting point for the analysis and design of telecommunication networks. The ultimate goal of telecommunications network service providers is to make customers (subscribers) happy while reducing network investments due to rising competition in the telecommunications services market.

It is mentioned earlier that the telecommunication network consists of different processing equipment's, interstate switching links and inter office trunks. The random nature of the call request makes difficulties in designing the equipments, switching links and trunks. Thus, the traffic analysis is the basic concern for the design of cost effective, efficient and effective architecture of networks. The success of a network can be assessed in terms of amount of carried traffic under normal or average loads and the frequency of the excess traffic with regards to the capacity of the network. The basic problem in the design of telecommunication networks involves the dimensioning of a route. To dimension a route correctly, we should have some information of its usage—that is, how many subscribers want to talk instantaneously over the route. The usage of a transmission route or a switch brings us into the dominion of traffic engineering.

In telecommunication system, traffic can be described as the usage of the server in the network. It is the total number of telephone calls over a group of circuits or trunks with regard to the duration of calls as well. The traffic flow is expressed as:

$$A = \lambda h \text{ Erlang}$$

where, A is the traffic intensity, λ is the call arrival rate and h is the average holding time.

The *call arrival rate*, λ is the average number of call request per unit time. The initiation of telephone-calls in any service area is random in nature. Therefore, a call request is a random variable. The holding time, h is the lasting time of the call or the average duration of possession of one or more traffic paths by calls.

There are two primary types of traffic-voice traffic and data traffic. For voice traffic, the calling rate can be defined as the number of calls per traffic path during the busiest hour. In a day, the 60 min interval in which the traffic gets the maximum value is called busy hour (BH). The busy hour may vary from exchange to exchange depending on the location and the community interest of the subscribers.

5.7.1   Trunking and Grade of Service

The telecommunication networks can handle a certain capacity at a specific "grade of service". Thus, it is essential to understand trunking theory to analysis and design of telecommunication network. The trunking concept accommodates a large number of users by sharing limited number of channels. The channel is assigned to each user from a pool of available channels, on demand. The user uses the channel on a per call basis, and after the execution of the call, the used channel is directly returned to the pool. The design for telephone switching center in a telecommunication system is based on the traffic intensity during the busy hour. The traffic generated by the subscribers sometimes exceeds the network capacity. There are two ways in which this overload traffic can be handled: the overload traffic can be declined without being serviced or held in a queue until the network facilities happen to be available. In the first case, the calls are lost and in the second case the calls are delayed.

Grade of Service (GOS): GOS is a measure of the ability of a user to access a trunked radio system during the busiest hour and can be expressed as the probability of call being blocked or delayed. Thus, to deal with GOS in traffic engineering, the clear concept of blocking criteria, delay criteria and congestion are essential.

5.7.1.1   Traffic Model

Telecommunication Systems can be classified as blocking systems or delay systems based on the handling of overflow traffic.

a. Blocking System: If the system is analyzed based on the blocking model, it is said to be loss system. Blocking is occurred if the resource is not available during the call request. Blocking criteria are often used for the dimensioning of switching networks and interoffice trunk groups.
b. Delay System: If the system is analyzed based on the queuing models due to the delay longer than the specific length of time (the delay probability), the system is called a waiting system or delay system. Delay criteria are used in telephone systems for the dimensioning of registers. In waiting system, the average call delay shall not exceed a given time.

5.7.1.2   Congestion

It is a condition in a switching center when a subscriber cannot obtain connection immediately to the called subscriber. In a circuit switching system, there will be a time of congestion during which no new calls can be accepted. There are two types of congestion.

1. Time congestion: It is the probability of blocking during which all servers are busy simultaneously.
2. Call congestion: It refers to the number of calls that fail to find a free server at first attempt. It is termed as loss system and known as the probability of loss while, in a delay system, it is referred to as the probability of waiting.

If the number of call sources is equal to the number of servers, the time congestion is finite, but the call congestion is zero. When the number of call sources is larger in compare to the servers, the probability of a new call arising is independent of the number already in progress and, therefore, the call congestion is equal to the time congestion. In general, time and call congestions are different but in most practical cases, the discrepancies seems to be small.

5.7.1.3   Erlang B formula

C = the number of trunked channels offered by a trunked radio system
A = the total offered traffic

The probability that a call is blocked is

$$P_b = \frac{\frac{A^c}{C!}}{\sum\limits_{k=0}^{C} \frac{A^k}{k!}}$$

One Erlang represents the amount of traffic intensity carried by a channel that is completely occupied (Table 5.1).

Problem 5.1

Over a 20-minutes observation interval, 40 subscribers initiate calls. Total duration of the calls is 4800 s. Calculate the load offered to the network by the subscribers and the average subscriber traffic.

Solution

Mean arrival rate, $\lambda = 40/20 = 2$ calls/min
    Mean holding time, $t_h = \frac{4800}{40 \times 60} = 2$ min/call
    Therefore,
    Offered load $= 2 \times 2 = 4$ E
    Average subscriber traffic $= 4/40 = 0.1$ E

Table 5.1 Network capacity planning blocking probability, GoS

n	0.5%	1.0%	2.0%	3.0%	5.0%	10.0%	20%
1	0.01	0.01	0.02	0.03	0.05	0.11	0.25
2	0.11	0.15	0.22	0.28	0.38	0.60	1.00
3	0.35	0.46	0.60	0.72	0.90	1.27	1.93
4	0.70	0.87	1.09	1.26	1.52	2.05	2.95
5	1.13	1.36	1.66	1.88	2.22	2.88	4.01
6	1.62	1.91	2.28	2.54	2.96	3.76	5.11
7	2.16	2.50	2.94	3.25	3.74	4.67	6.23
8	2.73	3.13	3.63	3.99	4.54	5.60	7.37
9	3.33	3.78	4.34	4.75	5.37	6.55	8.53
10	3.96	4.46	5.08	5.53	6.22	7.51	9.69
12	5.28	5.88	6.61	7.14	7.95	9.47	12.0
15	7.38	8.11	9.01	9.65	10.6	12.5	15.6
20	11.1	12.0	13.2	14.0	15.3	17.6	21.6
25	15.0	16.1	17.5	18.5	20	22.8	27.7
30	19	20.3	21.9	23.1	24.8	28.1	33.8
35	23.2	24.6	26.4	27.7	29.7	33.4	40.0
40	27.4	29.0	31.0	32.4	34.6	38.8	46.2
45	31.7	33.4	35.6	37.2	39.6	44.2	52.3
50	36	37.9	40.3	41.9	44.5	49.6	58.5

5.8   Sample Questions

1. What is meant by telecommunication network? State the major elements of it.
2. What is on hook and off hook? When a subscriber goes "off hook" condition.
3. How the switching systems can be classified?
4. Draw the pulse dialing waveform for the number 414.
5. Sketch and explain the typical hierarchical network architecture.
6. With neat diagram, illustrate the classification of switching system.
7. Explain the international telephone numbering format.
8. Assume each user of a system averages three calls per hour, each call lasting an average of 5 min. What is the traffic intensity for each user?
9. Find the number of users that could use the system with 1% blocking if only one channel is available.
10. What is the maximum system capacity in Erlangs when providing a 2% blocking probability with four channels, with 20 channels, with 40 channels?
11. During a busy hour, 1400 calls were offered to a group of trunks and 14 calls were lost. The average call duration has 3 min. Find

    (a) Traffic offered
    (b) Traffic carried
    (c) GOS and
    (d) The total duration of period of congestion.

References

1. T. Viswanathan, *Telecommunication Switching Systems and Networks* (Prentice hall)
2. R.L. Freeman, *Fundamental of Telecommunications* (Wiley, New York, 1999)
3. R.L. Freeman, *Telecommunication System Engineering*, 4th edn. (Wiley-Interscience, 2004)
4. T. Anttalainen, *Introduction to Telecommunications Network Engineering*, 2nd edn. (Artech House Publishers, March 2003)

Chapter 6
Cellular Telephony System

6.1 Introduction

Cellular mobile radio communication is a significant movement in the provision of recent communication services. It is extremely popular and most lucrative communication worldwide. The aim of cellular phone is to provide telephone services to subscribers while in motion. In early days, the mobile radio systems use a single, high power transceiver with an antenna mounted high on a tall tower to obtain a large coverage area. Though, it could achieve large coverage but couldn't support large number of subscribers as the regulatory agencies could not able to allocate spectrum as required with the growing demand for mobile services. Therefore, it became authoritative to reform the radio telephony system to accomplish large capacity with efficient use of the limited radio spectrum while covering very large areas. The cellular concept can solve both the issues-overcrowding spectral and user capacity. It offered very high capacity by replacing a single, high power transceiver (large cell) with many low power transceivers (small cells) using limited spectrum. Each cell provides coverage of small portion of the total service area. The total number of available channels for the entire system is distributed among a group of cells called cluster in which neighboring cell uses different sets of channels. This ensures that the interference between base stations and the mobile users under their control is minimal.

6.2 Basic Cellular Telephony

A cellular telephony system is linked to the public switched telephone network (PSTN) or other networks through gateway mobile switching centre (MSC). The basic cellular architecture shown in Fig. 6.1 consists of mobile stations, base stations and a mobile switching centre (MSC). The MSC is occasionally called a

© The Author(s) 2018   89
M. A. Matin, *Communication Systems for Electrical Engineers*,
SpringerBriefs in Electrical and Computer Engineering,
https://doi.org/10.1007/978-3-319-70129-5_6

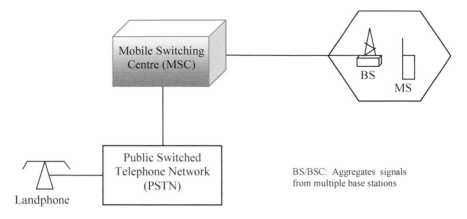

Fig. 6.1 Basic architecture of cellular system

mobile telephone switching office (MTSO), due to its responsibility for linking all mobiles to the PSTN in a cellular system. The mobile station have a transceiver, an antenna, control circuitry and is used a portable hand-held unit. The mobile station is connected to the base station through air-link. The base station is equipped with several transmitters and receivers and is placed at the center or on the edge of a cell. It can handle full duplex communications and usually have towers where antenna mast is often mounted. In a typical network arrangement, the many base transceiver stations (BTS) that are spread across a city are then linked to a centralized base-station controller (BSC). The link known as backhaul connects the BSC to MSC. The MSC handles the activities of all the base stations such as keeps track of users, manages incoming and outgoing calls and controls the hand-off between mobile phones and the base stations.

6.3   Cellular Telephony Evolution

The evolution of cellular telephony is depicted in Fig. 6.2. The first generation (1G) refers to analog cellular technologies which became available in 1980's. It was relied entirely on FDMA/FDD and analog FM. The Second generation (2G) wireless systems were first brought in the market in early 1990s which is the upgradation of the first generation of analog cellular systems. 2G standards use digital modulation formats and TDMA/FDD and CDMA/FDD multiple access techniques. Global system for mobile communication (GSM), Interim standard-136 (IS-136), and Interim standard-95 (IS-95) are the different standards of 2G cellular technologies. 2G standards are improved over time to include data communications, first with circuit switched data transport via high speed circuit switched data (HSCSD), then packet data transport via general packet radio services (GPRS) and Enhanced data rate for GSM Evolution (EDGE). Further development or

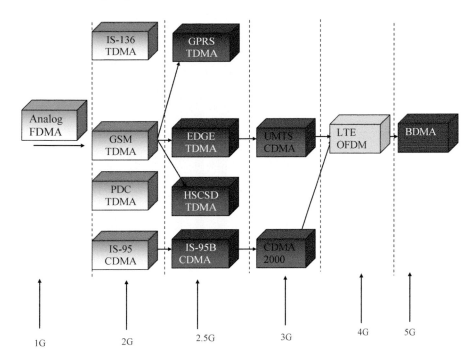

Fig. 6.2 Evolution for cellular telephony

improvements is done using 3G standard (UMTS) and fourth generation (4G) LTE Advanced standards. The development is further progressing due to exponential increase in demand of the users (shown in Fig. 6.3) and therefore, 4G will now be replaced with 5G.

6.4 Cellular Concept

In cellular communications, a region is geographically divided into cells. Each cell has its own low power base stations for transmission covering a limited area. The radio channels are then assigned to these cells in an intelligent way in order to maintain adequate call performance while keeping the interference at low level, and provide services to these areas. In theory, the cells are hexagons in shape, but in practice, they are less regular.

In cellular system, the antennas in the base stations are designed to cover its designated area within the cell. The same set of channels is used in different cells which are away from one another with an adequate distance to maintain the interference at tolerable limit. The neighboring cells cannot use the same set of frequencies in order to avoid interference. The design process of choosing and

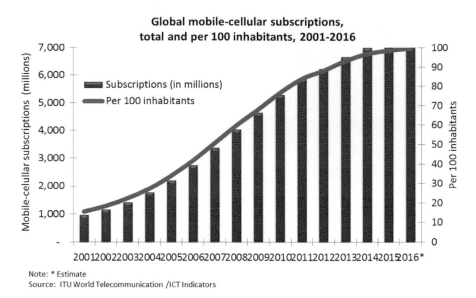

Fig. 6.3 Exponential increase in cellular subscription

assigning set of channels to all cellular base stations within a system is called
frequency reuse or frequency planning.

6.4.1 Frequency Reuse

Frequency reuse is the backbone of cellular concept. It implies that same set frequencies is used in different cells within different clusters in a given coverage area.
Figure 6.4 illustrates the concept of frequency reuse. The cells in a cluster are
labeled with the different letters to represent different group of channels which are
used again and again to cover the whole geographical area. The cell structure is

Fig. 6.4 Illustration of
frequency reuse concept

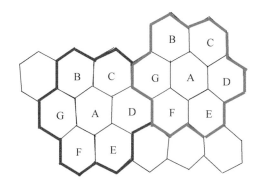

hexagonal shown in Fig. 6.4. This is a basic structure of the radio coverage for each base station, but commonly accepted since the hexagon allows easy and convenient analysis of a cellular system. The footprint is the actual radio coverage of a cell and it is estimated from field measurements or propagation prediction models.

To demonstrate the frequency reuse concept, S duplex channels is considered in a cellular network where the S channels are distributed among N cells. Each cell has assigned a group of k channels (k < S), (i.e. equal number of channels), then

$$S = kN \tag{6.1}$$

The N cells, which combinedly exploit the total number of available channel is called a cluster. As the cluster is repeated M times to cover the entire geographical area, the total capacity of the system, C, can be expressed as

$$C = MkN = MS \tag{6.2}$$

If the total geographical area is A_{total} where each cell cover a area of A_{cell}, then the total number of required cells is

$$M \times N = \frac{A_{total}}{A_{cell}} \tag{6.3}$$

The Eq. (6.2) demonstrates that the capacity of a cellular network is directly proportional to the number of repeated clusters in a specific service area. The value of the cluster size N is usually equal to 4, 7, or 12. One way to increase the capacity is to reduce cluster size N, maintaining the cell size constant because more clusters are needed in this case to cover this given area, and hence get more capacity. A large cluster size means that the co-channel cell distance is larger. On the contrary, a small cluster size means the distance is less between co-channel cells. As a result, the value of N is determined based on the tolerable interference between a mobile or base station while retaining an adequate quality of communications.

Problem 6.1

Assume the FDD cellular telephony system has a total bandwidth of 33 MHz, uses two 25 kHz simplex channels to offer radio transmission and reception using full duplex voice and control channels. Calculate the number of channels per cell if $N = 4, 7, 12$.

Solution

Total bandwidth of FDD cellular system = 33 MHz
Full duplex channel bandwidth = 25 kHz × 2 = 50 kHz
Total number of available channels = 33 MHz/50 kHz = 660

N = 4, Channel per cell = 660/4 = 165 channels
N = 7, Channel per cell = 660/7 = 95 channels
N = 12, Channel per cell = 660/12 = 55 channels

6.5   Hierarchical Cell Structure

There are few reasons behind using a hierarchical cellular infrastructure to support cells with different sizes which are as follows:

- *To broaden the coverage area that is difficult to reach.*
- *To enhance the capacity of the system for densely populated areas.*
- *Occasionally, an application requires specific coverage.*

In the recent deployment of cellular network, different cell sizes are used to offer a comprehensive coverage to support a range of applications. Dividing the cells into a hierarchy allows the network to effectively use the geographical area and provide efficient services.

Femtocells: Smallest cells in size and are employed for connecting personal equipment such as laptops, notepads and cellular telephones. These cells cover short distance about few meters where all the devices are to be in the range of the user.

Picocells: These are small cells typically covering local indoor networks inside a building to add system capacity. The coverage of these networks is within the range of few tens of meters.

Microcells: These cells cover a limited area and uses power control to limit its coverage. The coverage area is less than two kilometer and is employed in urban areas to maintain PCS.

Macrocells: Macrocells are the conventional cells installed during the initial phases of the cellular network that generally provide radio coverage to the metropolitan areas. These cells cover areas of few kilometers and their antennas are mounted on a tower with a height above the typical buildings.

Megacells: The coverage of Megacells is nationwide areas in the range of hundreds of kilometers and is primarily used with satellites.

6.6   Handoff in Cellular System

As the mobile equipment travels around a cell, the signal to interference ratio of its received signal from the base station of the cell varies. If the signal to interference ratio (SIR) becomes worse than a predetermined level, the call will be dropped. During a call, if a mobile travels towards a neighboring cell or other cells, the MSC needs to hand over the call to a new channel belongs to the new base station. Otherwise, the will be dropped due to receiving low SIR from the serving base station. In practice, this is a sophisticated process which involves not only recognizing a new base station, but also relocating the channels belongs to the new base station without disruption. If the new base station does not have available channel, the call will be dropped. The following things need to be kept in mind while dealing calls in cellular system.

- Handoffs need to be processed in any cellular network. However, unnecessary handoffs should be avoided as Handoffs are expensive to execute.
- In most handoff strategies, the request for handoff gets precedence over call initiation requests.
- Effective and infrequent handoff must be executed as best as possible and also be unnoticeable to the users. Unreliable and ineffective handoff process will decrease the quality and reliability of the system.

To fulfill these requirements, we must identify an optimal signal level at which handoff can take place. Once, a specific signal level is determined for satisfactory voice quality at the BS receiver, a slight increase in signal level is used as a threshold for handoff.

This margin is given by, $\Delta = P_{r,handoff} - P_{r,minimum\,usable}$, which should be within acceptable range.

- If the value of Δ is too large, redundant handoffs will take place and this can be burdened the RNC.
- On the other hand, if the value of Δ is too small, inadequate time will get to hand over a call before it is lost due to weak signal.

Therefore, Δ is very important. Figure 6.5 demonstrate the case where a handoff is not happened due to poor signal which is below the minimum acceptable signal to be the channel active.

Figure 6.5 shows a mobile station (MS) is traveling from one base station (BS1) to a new base station (BS2). The average received signal strength from BS1 is diminishing when the MS is moving away from it. Correspondingly, the average signal strength rises if the MS comes close to base station. However, received signal from the new BS should be sufficiently stronger than the current one for making the handoff. The repeated handoff between two BSs could be occurred due to rapid signal variations from both base stations. The handoff is redundant if the serving BS is satisfactorily strong.

In first generation analog cellular systems, the handoff process typically takes about 10 s and the value of Δ is in between 6 and 12 dB. In second generation cellular systems such as GSM, the handoff procedure is mobile assisted. Each mobile station estimates the received power from surrounding base station and continuously reports to the serving base station, once the decision is made, handoff is initiated which typically requires only 1 or 2 s.

As of the user's perspective, the call drop in the middle of a conversation is more irritating than denying a new call request. To enhance the QoS, handoff request is given precedence rather than initiating a call before allocating voice channels. The guard channel concept is one of the methods for giving priority to handoffs. In this method, few channels of a cell are reserved solely for serving the necessities of handoff from enduring calls. This method reduces the total available channels for originating calls.

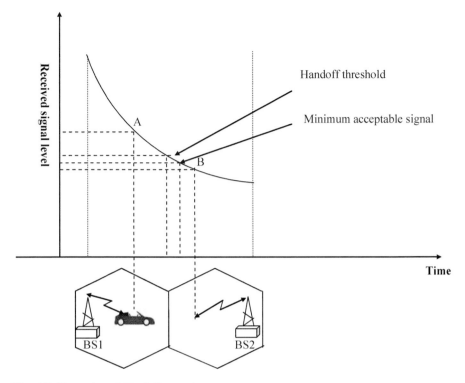

Fig. 6.5 Illustration of Handoff scenario

6.6.1 Types of Handoffs

There are basic two types of handoff techniques-hard and soft handoffs. Typically, the hard handoff technique is of two different types-intra and intercell handoffs. On the other hand, the soft handoff is of two other types-multiway soft handoffs and softer handoffs.

- Hard Handoff

The network selects whether a handover is needed or not based on the signal strengths of the current channel, and the strengths of broadcasting channels of neighboring cells. If required, the connection between the current NodeB and the UE is temporarily broken and a new connection is created between the new NodeB and the UE. The break might take place so quickly but there exist a moment of disruption to the call.

- Soft Handoff

Soft handoff is "make before break". It occurs while the user equipment (UE) is traveling in the overlapping region of NodeB's. The UE sends and receives the signals from the two NodeB's. The rake receiver is required in the UE to add up the signals from two NodeB's. In soft handoff, the UE receives outer loop power control bits from the two NodeB's and perform operation to decide how to adjust its power.

- Softer Handover

Softer handover is specific case of soft handover where the radio links that are added and removed are from two sectors but under the same Node B.

6.6.2 Practical Handoff Scenarios

6.6.2.1 Umbrella Cell Approach

In practice, several issues come up while going to design cellular systems for a large range of mobile speeds. High speed users can pass through the cell designated area very quickly, whereas low speed users do not require a handoff during a call. The other practical issue is to get new cell sites due to zoning laws, ordinances, and other non-technical barriers but additional capacity is provided through additional cell sites. To handle this issue, cellular service provider can accommodate more channels and base stations at the identical location on the same tower or building of an existing cell, before finding new site locations. The antennas mounted at different heights on the identical building or tower with different power levels can offer "large" and "small" cells coverage that are confined at the same location. This phenomenon is known as umbrella cell approach and is used to cover large area to high speed users whereas small area is covered to users moving at low speeds. The advantages of umbrella cell approach are:

- *the minimum number of handoffs for high speed moving subscribers and*
- *additional microcell channels for low speed subscribers.*

6.6.2.2 Cell Dragging

One of the realistic problems during handoff in cellular networks is cell dragging. This phenomenon is resulted from low speed users that receive reasonable signal strength from the previous base station. It happens in an urban area due to the existence of line-of-sight (LOS) path between the mobile user and the base station. This is due to the movement of the user from the base station at low speed which results in good average signal strength. Therefore, handoff is not made even when the user has traveled to the new cell and received signal above the handoff threshold from the previous cell. This generates a probable interference and traffic management problem, as the user is moving deep into a neighboring cell. To resolve the cell dragging problem, the parameters for handoff thresholds and radio coverage need to be tuned with great care.

6.7 Interference and System Capacity

Though frequency reuse boosts spectral efficiency, it also brings interference, which affects the performance of cellular radio systems. The interference on voice channels causes cross talk, where the user can listen noise in the background. On the other hand, interference on control channels creates error in signaling which leads to missed and blocked calls. Interference is being identified as a key bottleneck in increasing capacity and is often blamed for dropping calls. The two main types of interference are co-channel interference and adjacent channel interference. Co-channel interference cannot be remedied by simply boost up energy of a transmitter. This is due to increased interference level to neighboring co-channel cells while increases transmit power. In order to reduce co-channel interference, the cells with same set of frequencies (co-channel cells) must be placed at a minimum distance to provide adequate isolation among them (Fig. 6.6). Since co-channel interference is depending on shadowing and multipath fading, the system designer considers the worst-case propagation conditions in determining the co-channel distance (Table 6.1).

Let i_0 is the number of co-channel interfering cells. Then the signal to interference ratio (SIR) for a mobile receiver during monitoring a forward channel can be stated as

$$\frac{S}{I} = \frac{S}{\sum_{i=1}^{i_0} I_i} \tag{6.4}$$

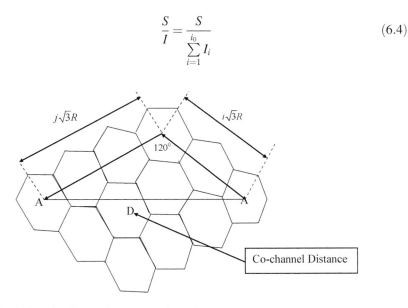

Fig. 6.6 Finding the distance between co-channels

Table 6.1 Co-channel reuse ratio for different values of cluster size

(i,j)	Cluster size $N = i^2 + ij + j^2$	Frequency reuse ratio $(1/N)$	Co-channel reuse ratio $(\sqrt{3N})$
(1,1)	3	0.33	3
(2,0)	4	0.25	3.46
(2,1)	7	0.14	4.58
(3,0)	9	0.11	5.2
(2,2)	12	0.083	6
(3,1)	13	0.077	6.245
(4,0)	16	0.0625	6.93
(3,2)	19	0.053	7.55

where, S is the signal strength received from the serving base station and I_i is the interfering power arises from the ith interfering co-channel cell base station. If D_i is the distance of the ith interfere from the mobile, then the interfering power at the given mobile receiver will be proportional to $(D_i)^{-n}$. If all the base stations have equal transmission and the path loss exponent is constant all over the serving area, then SIR for a mobile can be estimated as

$$\frac{S}{I} = \frac{R^{-n}}{\sum_{i=1}^{i_0}(D_i)^{-n}} \tag{6.5}$$

Considering the similar cell shapes and the location of mobile equipment is at the cell edge, it is observed for a seven-cell cluster that the mobile is at a distance $D - R$, D, $D + R$ from the surrounded interfering cells in the first tier (Fig. 6.7).

To ease of SIR calculation, here the first tier interfering cells are considered. If the interfering base stations are at the same distance from the given mobile station which is D, then the above equation can be written as

$$\frac{S}{I} = \frac{(D/R)^n}{i_0} = \frac{(\sqrt{3N})^n}{i_0} \tag{6.6}$$

It is noted here that Eq. (6.6) is based on the hexagonal cell structure where the distance from all interfering cells to the given mobile receiver, are about equal and therefore presents an approximate result.

Problem 6.2
To have acceptable forward channel performance of a cellular system, the required signal-to-interference ratio is 15 dB. Calculate the frequency reuse factor and cluster size that is used to achieve maximum capacity if the path loss exponent is

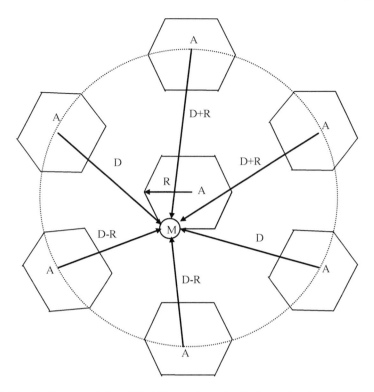

Fig. 6.7 Graphical representation of first tier co-channel cells for cluster size of seven

(a) $n = 4$ (b) $n = 3$. Consider six co-channel cells are in the first tier, and they are all equidistant from the mobile station. Use appropriate approximations.

Solution

(a) $n = 4$

First, assume a seven cell reuse pattern. Using the equation for signal to interference ratio (SIR), we get

$$\frac{S}{I} = \left(\frac{1}{6}\right) \cdot \left(\sqrt{3 \times 7}\right)^4 = 75.3 = 18.66 \, \text{dB}$$

Since this is above the minimum SIR requirement, $N = 7$ can be used. The frequency reuse factor can be calculated using the equation below:

$$\rho = 1/N = 1/7$$

(b) $n = 3$,

First, assume a seven cell reuse pattern. Using the equation for signal to interference ratio (SIR), we get

$$\frac{S}{I} = \left(\frac{1}{6}\right) \cdot \left(\sqrt{3 \times 7}\right)^3 = 16.04 = 12.05\,\text{dB}$$

As the calculated SIR value is below the minimum required SIR, we require a higher value of N. The next possible value of $N = 12$, and the corresponding SIR ratio is,

$$\frac{S}{I} = \left(\frac{1}{6}\right) \cdot \left(\sqrt{3.12}\right)^3 = 36 = 15.56\,\text{dB}$$

As this is above the minimum required SIR, $N = 12$ is used. The frequency reuse factor can be calculated using the equation below:

$$\rho = 1/N = 1/12$$

Problem 6.3
The area of certain city is 1300 square miles, covered by a cellular system with a cluster size of seven. The radius of the cell is four miles. The spectrum of 40 MHz is distributed within the city with a full duplex channel bandwidth of 60 kHz. If the traffic intensity in each cell is 84 Erlangs and the offered traffic per subscriber is 0.03 Erlangs, calculate the followings:

(a) Number of cells in the coverage area
(b) Number of channels in each cell
(c) Maximum carried traffic
(d) Total number of subscribers
(e) Number of mobiles per unique channel
(f) Theoretical maximum number of users served at one time by the system.

Solution

The total area of the city = 1300 miles, and cell radius = 4 miles
    The cell area of hexagonal shape can be calculated using the following figure (Fig. 6.8).

(a) Cell area = $2.5981\,R^2$, thus the coverage of each cell is $2.5981 \times (4)^2 = 41.57$ square miles
    Hence, the number of cells in the service area is, $N_c = 1300/41.57 = 31$ cells
(b) The total number of channels in each cell = allocated spectrum/(channel bandwidth $\times$ cluster size) = 40,000,000/60,000 $\times$ 7) = 95 channels/cell
(c) Maximum carried traffic = total cells $\times$ each cell traffic intensity = 31 $\times$ 84 = 2604 Erlangs

Fig. 6.8 Hexagonal cell area

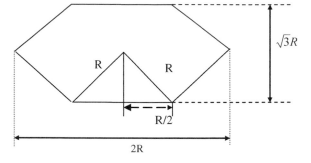

(d) Each user traffic = 0.03 Erlangs
     Total users = total traffic/traffic per user = 2604/0.03 = 86,800 users
(e) Number of mobiles per channel = number of users/number of channels =
     86,800/666 = 130 mobiles/channel
(f) The theoretical maximum number of subscribers is the number of available
     channels in the system (all channels occupied) = $C \times N_c = 95 \times 31 = 2945$,
     which is 3.4% of the customer base.

6.8   Techniques for Improving Coverage and Capacity

The wireless cellular telephony is quickly replacing the use of wired telephone
network and thus the demand for it, increases at remarkable pace. Therefore, the
number of channels allocated to a cell ultimately becomes inadequate to serve the
required number of users. Currently, few approaches are adopted to accomplish the
additional channels per unit service area. These techniques include cell splitting,
sectoring, and microcell zone concept, in practice any method to reduce interference
will increase the system capacity. Cell splitting provides a systematic expansion of
the cellular system. On the other hand, sectoring exploits directional antennas to get
additional control the interference to improve signal to interference ratio (SIR) and
frequency reuse of channels. The zone microcell concept divided the cell into
multiple zones and extends the cell boundary to hard-to-reach places.

6.8.1   Cell Splitting

Cell splitting technique does rescaling the system to achieve improved capacity by
shrinking the cell size while maintaining the co-channel reuse ratio D/R constant. It
enhances the number of channels per unit area.
    Cell splitting is the method of partitioning a crowded cell into microcells; each
microcell has its own base station with reduced antenna height and transceiver

Fig. 6.9 Illustration of cell splitting

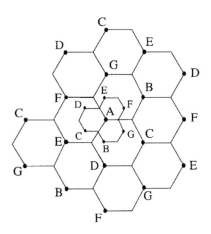

power. Cell splitting provides increased capacity of a cellular system since it increases the number of times that channels are reused. The microcells having smaller radius than the original cells are placed between the existing cells which increases capacity caused by the additional channels per unit area.

Cell splitting phenomenon is shown in Fig. 6.9 where the base stations are located at corners of the cells. Assume heavy traffic load is experienced in the region served by base station A. As a result, cell splitting is required for additional number of channels in the reduced area to accommodate traffic loading. It is noted in the figure that the original base station A is being enclosed by six new microcells. These microcells are created in intelligent way without violating frequency reuse constraint. For instance, the microcell base station labeled C is located half way between two larger cells base station using the same set of frequencies C. This also happens for the other microcells in the Fig. 6.9. Thus, cell splitting process provides additional number of channels through creating the new microcells.

6.8.2 Sectoring

In sectoring, the cell radius is kept constant and seeks techniques to reduce the D/R ratio. Since sectoring technique improves SIR, the cluster size can be reduced. Each cell is divided into radial sectors with directional base station antennas to improve the performance against co-channel interference. Since SIR is improved, larger capacity can be achieved by shrinking the cluster size, that means reduce the number of cells.

The method for reducing co-channel interference and hence increasing system capacity by using directional antennas at the base station is called sectoring. The reduction of the co-channel interference relies on the amount of sectoring used. Usually, cell is divided into three 120° sectors or six 60° sectors as shown in

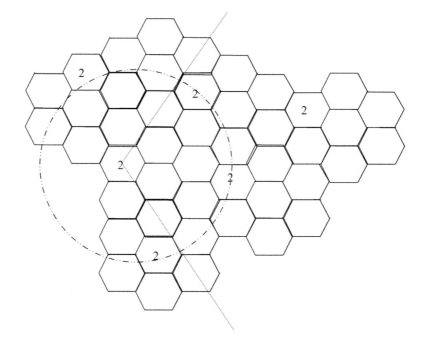

Fig. 6.10 Illustration of sectoring technique. Cell is partitioned into 120° to reduce number of interfering co-channel cells

Fig. 6.10. 120° cell sectoring reduces the interference by roughly a factor of three whereas 60° sectoring by a factor of six.

6.8.3   Microcell Zone Concept

The problem in sectoring is the increased number of handoffs which consequently increases the load on the switching and control link elements of the mobile system. The microcell zone concept is the solution for the above problem which is demonstrated in Fig. 6.11. In this concept, cell is divided into multiple zones and all zone sites (represented as Tx/Rx in Fig. 6.11) are connected through coaxial cable, fiber optic cable, or microwave link to a single base station and share the identical radio equipment. When a mobile user travels between microcell zones of the cell, it is served by simply switching the channel with the strongest signal. In this concept, the base station channel can be assigned to any zone through zone selector by the base station.

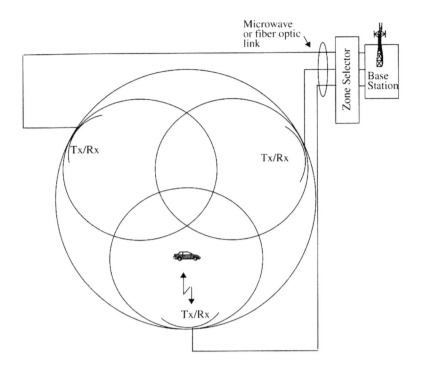

Fig. 6.11 The microcell zone concept

6.9 Sample Questions

1. Define the following terms:

    i. Handoff
    ii. Base Station.

2. Explain the concept of "frequency reuse" as applied to cellular Communications and derives an expression for the "signal to interference ratio" for an omni-directional system. Consider that all cells have equal radii, with the base stations of equal power and are positioned in the center of each cell.
3. Determine the "signal to interference ratio (SIR)" for the following reuse designs: $N = 1, N = 3, N = 4, N = 9$ and $N = 12$. Calculate the SIR contribution due to just the first tier where the path loss exponent is three.
4. Demonstrate that the co-channel reuse ratio is given by $\frac{D}{R} = \sqrt{3N}$ for a hexagonal cell structure.
5. If a 20 MHz of available spectrum is assigned for a FDD cellular system and each simplex channel has a bandwidth of 25 kHz, calculate

  (a) the number of duplex channels,
  (b) the total number of channels per cell site, if N = 4 cell reuse is used.
6. Mention different techniques used to increase the coverage and capacity of a
   cellular system.
7. Calculate the signal to interference ratio (SIR) for a nine cell reuse cluster with
   120° sectors. Assume the path loss exponent is four.

References

1. T.S. Rappaport, *Wireless Communications: Principles and Practice*, 2nd edn. (Prentice hall,
   2001)
2. D. Tse, P. Viswanath, *Fundamentals of Wireless Communication* (Cambridge University Press,
   2005)
3. W.C.Y. Lee, *Wireless and Cellular Communications*, 3rd edn. (McGRAW-HILL, 2006)
4. M.A. Matin, *Handbook of Research on Progressive Trends in Wireless Communications and
   Networking* (IGI Global, Hershey, PA, USA, 2014)

Chapter 7
Satellite Communication Systems

7.1 Introduction

In satellite communication (Fig. 7.1), satellites are used to relay signals between ground stations. The signal is received from one ground station and re-transmits to one or more receiving ground stations elsewhere with the help of satellite. In this process, signal from ground station is first sent towards the satellite, then amplifies the signal and sends it back to the other ground station. As the signal is transferred through space, this kind of communication is also known as space communication. The realization of satellite communications from an approach to practical has become true due to the technological breakthroughs that happened during and after the World War II. In October 1957, former Soviet Russia launched the first artificial satellite Sputnik-I in the low earth orbit and Clark's idea turned into reality in 1963 after successful launching the first geosynchronous satellite SYNCOM by NASA. The weight of Sputnik was 184 lbs and was placed in an orbit of 560 miles above the earth. It is transmitted on 20.005 and 40.002 MHz. Nevertheless, this spacecraft was much more than a technical success as it had a remarkable impact particularly in psychological and political point of view which results in a technological rivalry between the United States and Russia, long term planning. As a result, US Explorer-1 was initiated in January 1958 by a Jupiter rocket after the launching of Sputnik, and pushes forward the space race between Russia and US.

There are two types of satellites that are commonly used in satellite communication—one is active satellite and the other one is passive satellites. Passive satellites are effectively utilized in the early years of satellite communications and with the development in technology, active satellites have entirely replaced the passive satellites.

© The Author(s) 2018
M. A. Matin, *Communication Systems for Electrical Engineers*,
SpringerBriefs in Electrical and Computer Engineering,
https://doi.org/10.1007/978-3-319-70129-5_7

Fig. 7.1 Satellite
communication

7.2   Active and Passive Satellites

In active satellites, signal is amplified and retransmits to the earth. The transmission
can be made at the same time or buffering in the memory of the satellites. Some
active satellites also have programming and recording features which can be dis-
played and observed. In 1957, Russia launched the first active satellite. The signals
coming from the satellite are of very low intensity when reach the earth. Therefore,
amplification is necessary which is made by the receivers themselves and after
amplification these become available for further processing.

A typical satellite operational link involves an active satellite and two or more
earth stations where the earth station transmits the signal up to the satellite in one
frequency termed as the up-link frequency. The signal is then amplified in satellite,
converts it to the downlink frequency, and transmits it back to the ground. The
signal is next picked up by the receiving terminal. In 1958, US air force launched
the world's first active satellite SCORE at orbital height of 110–900 miles.

The passive satellites reflect the incident electromagnetic radiation based on
scattering phenomenon of electromagnetic waves from different surface areas. An
electromagnetic wave impinge on a passive satellite reflects signal towards one
earth station to other or from few earth stations to few others. In passive satellites,
the reflected signal is not amplified at the satellite and as a result, only a fraction of
transmitted signal is in fact arrives at the receiver. The passive satellites used at the
beginning of satellite communications were both artificial as well as natural. Moon
is a passive satellite for our planet. Though passive satellites were simple, the
communications between two remote places were established after overcoming
huge technical problems. The large path loss representing attenuation of the signal
is due to long distance between the transmitter and the receiver via the satellite. The
attenuation is one of the most severe problems in satellite communication.

7.3   Satellite Communications Spectrum

The satellite transmission bands of interest are the C-, Ku- and Ka-bands (Table 7.1). The C-band is the oldest allocation and uses around 6 GHz for transmission (uplink) and between 3.7 and 4.2 GHz for reception (downlink). Ku-band is the most prevalent transmission format for satellite TV in Europe and uses around 14 GHz for uplink and between 10.9 and 12.75 GHz for the downlink. Ka-band uses around 30 GHz up- and between 18 and 20 GHz downlink frequency. The highest frequency band typically gives access to the broader bandwidths, but is also more vulnerable to signal degradation due to 'rain fade'.

7.4   Orbits for Communication Satellites

Satellite that follows a gravitationally curved path around a planet or a star is called orbit. Typically, satellite orbit is elliptical in shape where the planet is located at one of the two foci of the ellipse. The circular orbit is just as an ellipse whose foci concur at the center of the circle.

   In early days, satellites are placed in low earth orbits (160–2000 km) due to the technical barrier of launching vehicles to place satellites in higher orbits. Simple launch vehicles are used to place huge mass satellites into low earth orbit. With the advancement of launch vehicles and satellite technologies, most of the satellites for telecommunications use geostationary orbit because of its huge advantages. Most videos or TV communication systems use geostationary satellites. All geostationary orbits have a semi-major axis of 42,164 km. Satellite orbits can be classified into three based on the inclination of the orbital plane.

- Polar Orbit
- Equatorial Orbit
- Inclined Orbit.

Table 7.1   Satellite bandwidth and applications

Band	Frequency		Bandwidth	Applications
VHF	<200 MHz			Messaging
UHF	200 MHz–1 GHz			Messaging and positioning, voice and fax
L	2/1 GHz			Mobile Satellite Services (MSS)
S	4/2.5 GHz			Fixed Satellite Services (FSS)
	Frequency, GHz			
	Uplink	Downlink		
C	5.9–6.4	3.7–4.2	500 MHz	Fixed point to point ground station
X	7.9–8.4	7.25–7.75	500 MHz	Mobile, radio relay, military only
Ku	14–14.5	11.7–12.2	500 MHz	Broadcast, fixed point service, non-military
Ka	27.5–31	17.7–21.2	3.5 GHz	

Fig. 7.2 Satellite Orbits with
respect to equatorial plane

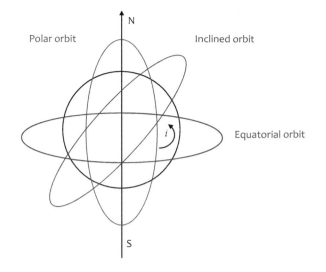

A satellite whose orbiting has an inclination of 90° or near 90° with the equatorial plane is in a polar orbit. If a satellite orbital plane is parallel to the equatorial plane of the earth, it is in an equatorial orbit and if not identical with the equatorial plane, it is in an inclined orbit. It is the angular distance of the orbital plane from the reference equatorial plane. The inclination of the orbit determines the satellite covered path and the greater the inclination, the greater the amount of the satellite covered surface area (Fig. 7.2).

Based on orbital height, satellite orbits can be also classified as (Fig. 7.3):

- Low Earth Orbit (LEO)
- Medium Earth Orbit (MEO)/Intermediate Circular Orbit (ICO)
- Highly Elliptical Orbit (HEO)
- Geosynchronous Earth Orbit (GEO).

Satellite orbits with altitudes from 160 to 2000 km are known as low earth orbit (LEO). LEO's in general circular in shape. Satellite orbits with orbital heights in the range of 2000 km to just below geosynchronous orbit at 35,786 km are known as medium earth orbit (MEO) or intermediate circular orbit (ICO). MEO is not limited to circular orbits. The higher elliptical orbit (HEO) satellites are located in the higher latitudes for communications. Russian Molniya satellites (launched on 23 April, 1965) are placed in eccentric elliptical orbits with a perigee of about 1000 km, apogee of 40,000 km, inclination of 63.435° and orbital period of around 12 h. In case of geosynchronous earth orbit (GEO), the satellite is placed in equatorial circular orbit with an orbital period of 24 h. The satellite appears stationary, always at the same point in the sky. Three satellites in GEO placed 120° spaced over equator provide communication services to the most part of the world.

Fig. 7.3 Satellite Orbits
(altitude classification)

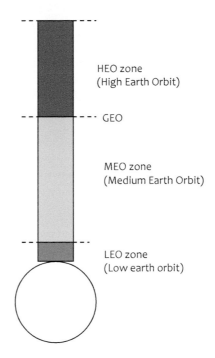

HEO zone
(High Earth Orbit)

GEO

MEO zone
(Medium Earth Orbit)

LEO zone
(Low earth orbit)

The altitude, h of GEO satellite is easily derived by balancing the two major forces acting on the satellite, viz.

$$mr\omega^2 = \frac{MmG}{r^2}, \tag{7.1}$$

where,

$M =$ mass of the earth, 5.98×10^{24} kg
$m =$ mass of the satellite
$r =$ distance of the satellite from the centre of the earth
$G =$ gravitational constant, $6.67 \times 10^{-11} \mathrm{Nm^2/kg^2}$
$\omega =$ angular velocity $= \frac{2\pi}{T} = 7.27 \times 10^{-5}$ rad/s

Substituting the known values in Eq. (7.1), we have

$$r = \left(\frac{6.67 \times 10^{-11} \times 5.98 \times 10^{24}}{7.27 \times 10^{-5} \times 7.27 \times 10^{-5}}\right)^{\frac{1}{3}} = 42,200 \,\mathrm{km}$$

Now, the radius of the earth, R, is 6400 km. Therefore, the altitude h of the satellite is given as $h = r - R = 35,800$ km.

7.5   Elements of Satellite Communications System

The two major elements of the satellite communications system are the *space segment* including the satellite itself, means for launching satellite, and satellite control centre and the *ground segment* with earth stations, communication links, user terminals and interfaces and network control centre. The earth stations are most often connected to the end users terminal through the terrestrial network or in case of small station connected to the end user terminal directly. The function of the earth station is to transmit a signal to the satellite and receive a signal from the satellite. The block diagram of satellite communication system is shown in Figs. 7.4 and 7.5.

7.5.1   Space Segment

The space segment (Fig. 7.6) comprises of satellite and the equipment on board of the satellite. The satellite consists of the payload and the subsystems. The payload consists of the receiving and transmitting antennas and all the electronic equipment which holds the transmission of the carriers. Communication satellites are complex and expensive to acquire and launch. The Placement of a satellite into the orbit and operating it for 12 or more years involves a great deal. Placement in orbit is accomplished by contracting with a spacecraft manufacturer and launch agency and allowing them about three years to design, construct, and launch the satellite. The mass and volume of the electric power supply of a satellite poses one of the most

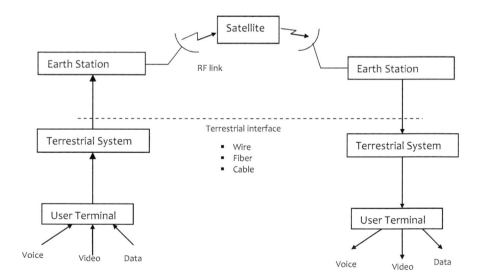

Fig. 7.4 Elements of satellite communication system

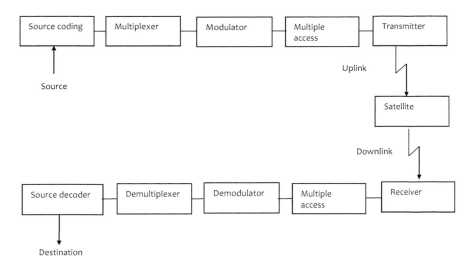

Fig. 7.5 Signal processing elements in satellite communications

restricting problems. In some applications such as television broadcasting or mobile and personal communications, the required electric power is over 10 kW which results in large weight and volume for communications satellites. Therefore, lightweight material is used for optimal design to support the load, withstand vibration and to keep the size and weight of the satellite small. Large temperature cycles are opted for the structure of the satellite.

Once the satellite is placed in the proper orbit, it becomes the responsibility of a satellite operator to control the satellite for the duration of its mission. It is fairly complex task and involves both sophisticated ground-based facilities as well as highly trained technical personnel. The tracking, telemetry, and command (TT&C) station (or stations) establishes a control and monitoring link with the satellite. Precise tracking data are collected periodically via the ground antenna to allow the pinpointing of the satellite's position and the planning of on-orbit position corrections. That is because any orbit trends to distort and shift with respect to a fixed point in space due to varying gravitational forces from the nonspherical earth and the pull of the sun and moon.

Fig. 7.6 Space segment

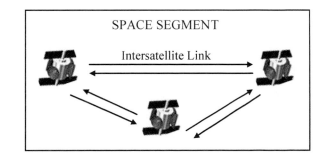

During the entire life time of 12–15 years, the communication satellite can earn revenue, to recuperate the initial and operating costs. As the satellite provides services over a long period of time in the space, the subsystems of the satellite are necessitated to be reliable for operation. In case of failure or ends of lifetime of a satellite, it is ceased. The lifetime is stated as the capability to maintain the satellite on station in the nominal attitude, and depends on the amount of fuel available for the propulsion system and altitude and orbit control.

7.5.1.1 Satellite Payload

The payload in most commercial communication satellites consists of two distinct parts—the transponder and the antennas.

(a) *Transponder*

There are two basic types of satellite transponders: repeater or bent pipe and processing or regenerative. In the repeater type, communication transponder captures the signals from earth, merely amplifies and frequency shifts the signals, whereas, in processing transponder, the RF carrier is demodulated to baseband signal and process the signals and later re-modulated the baseband signal in addition to frequency translation and amplification. Analog communication systems are entirely repeater type. Digital communication system may use any type. Figure 7.7a, b show the schematic diagrams of repeater type and regenerative type transponders respectively.

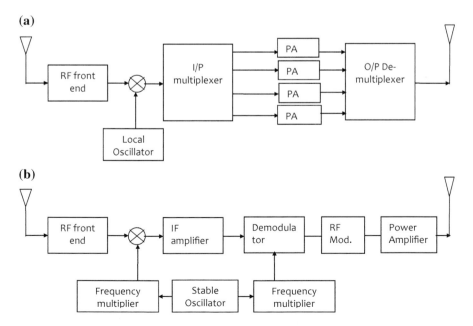

Fig. 7.7 a Repeater transponder. **b** Regenerative transponder

(b) *Satellite antennas*

The Satellite station receives and retransmits the signals using satellite antennas which maintain the link between the ground segment and the communications transponder. The size and shape of the satellite antennas depend on the specific coverage requirements of the system. Satellite antennas generate small multiple spot beams and increase the EIRP for the same HPA power. Each beam covers a certain portion on the earth surface called footprint. The gain of the satellite antennas is same both in transmission as well as reception because the generated beamwidths are same. This necessitates two separate antennas on the satellite but the diameters of the antennas will not be the same (depends on uplink and downlink frequencies). The beamwidth of a satellite antenna is equal to or less than the angle of view of the earth from the satellite, which is $17.5°$ in case of a geostationary satellite.

The core functionalities of satellite antennas are as follows:

- to capture the radio waves, that is transmitted from ground station at uplink frequency band and with a given polarization; to capture the desirable signals and reject undesired signals as possible;
- to transmit radio waves, in a given downlink frequency band and with a given polarization, to a specific region on the earth surface;
- to transmit more focused power i.e., reduce wastage of power.

7.5.1.2 Satellite Subsystems

The major subsystems of a satellite are:

- Altitude and orbit control system
- Propulsion System
- Electrical Power System
- Telemetry, Tracking and Command (TTC) System
- Thermal Control System.

Altitude and orbit control system: A satellite in orbit experiences small disturbances due to some weak forces such as attraction from moon, sun and other planetary bodies, pressure of solar radiation and variations in the earth's gravitational force. These forces cause the satellite to drift from its orbit, retard or accelerate its motion and affect its orientation towards the earth. Altitude and orbit control system maintains the orbital position of the satellite and different sensors and firing small thrusters are located in different sides of the satellite to control the altitude of the satellite.

Propulsion System: The primary function of the propulsion system is to produce forces which act on the centre of mass of the satellite. These forces adjust the satellite orbit, either to make sure the injection into a predestined orbit or to control drift of the nominal orbit. The propulsion system also provides torques to support the attitude control system. The forces caused by the propulsion units are reaction forces resulting from the expulsion of material. In the satellite, liquid fuel and oxidizer are carried as part of the propulsion system for firing the thrusters in order to hold on to the satellite attitude and orbit. The quantity of fuel and oxidizer also determines the effective life of the satellite.

Electrical power System: The electrical power in the satellite is obtained mostly from the solar panel. The beginning-of-life power output of the solar panel can be achieved with reasonably good accuracy, although precise on-ground measurements are typically not possible beyond solar cell string level. In orbit, the solar cells in the panel are exposed to the space radiation environment, which consists of free electrons, protons, and UV energy. Most of this charged flux emanates from the sun and follows the 11-year solar cycle. Another source of degradation is surface contamination from out gassing of spacecraft materials that deposit on the panels. The communication payloads and all other electrical subsystems in the satellite use this electrical power. Rechargeable battery is also used for providing electrical power during ellipse of the satellite.

Telemetry, Tracking and Command (TTC) System: Telemetry, Tracking and Command (TTC) system of the satellite station works together with its counterpart placed in the satellite control ground station. It allows ground operations to monitor and control the entire satellite. TTC deals with receiving control signals from the ground to initiate manoeuvres and to change the state or mode of operation of equipment. The telemetry system gets data from sensors on board the satellite and transmits these data through telemetry link to the satellite control centre which keeps an eye on the health of the satellite. TTC requirements must be considered during the design and optimization of the *payload*. Most satellites include TTC frequencies at the upper or lower edge of the communication band. However, this may not be appropriate if existing ground tracking stations cannot employ the same frequencies. The command system is used for switching on/off of different subsystems in the satellite based on the telemetry and tracking data.

Thermal control system: The purpose of thermal control is to maintain the temperature of different parts of satellite within the temperature ranges which facilitate it to operate satisfactorily, by offering its nominal performance, and defending any irreparable deterioration subsystems from the extreme temperature conditions of the outer space when it is not operating.

7.5.2   Ground Segment

The ground segment of satellite communication system supports the space segment and has a command (uplink) function as well as data (telemetry and mission function). It sets up the communications links between the satellite and the user. In large and medium system, the terrestrial microwave link interfaces with the user and the earth station. However, in the case of small systems, the end user's terminal interface is located at the earth station. The earth station (Fig. 7.8) consists of

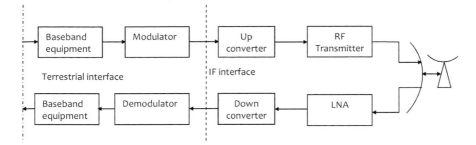

Fig. 7.8 Shows a schematic block diagram of an earth station

- Transmit equipment.
- Receive equipment.
- Antenna system.

In the ground station, the base band signal of terrestrial network is modulated on IF and then up-converted before transmitting to the satellite. The receiving ground station transmits the base band signal to the user directly or through the terrestrial link.

The baseband signals received at the earth stations are generally—telephone, television or data signals. It can relay information from a single source or from several sources through multiplexing of signals from the individual sources.

In early days, FM modulation scheme is used for analog voice and video signal transmission in satellite communications. Now, the transmission signal is mostly digital for both voice and video and the modulation schemes such as phase shift keying (PSK) and frequency shift keying (FSK) are implemented for transmission.

The monitoring, alarm and control equipment of the earth station has the following purposes:

- to monitor the network operations and controlling the station and managing the traffic;
- to give alarms in case of inaccurate operation or an occasion that affects the main station equipment or the link performance and allow detection of the equipment which is involved;
- to control the station equipment including adjustment of parameters, switching of redundant equipment and so on.

It should keep in mind that the correct functioning of network operations and control centre is necessary where the number of users in the network is large.

7.6   "Link Budget" for Satellite Communication

An accounting of signal strength and noise is an important part of system design and is known as "link budget". In this budget, the effective transmit power is calculated, losses are subtracted, noise is accounted for, and a link margin is determined (Fig. 7.9).

The Overall received power (in dB):

$$P_R = EIRP + G_R - Losses$$

$$P_R = EIRP + G_R - (FSL + RFL + AML + AA + PL) \tag{7.2}$$

Effective isotropic radiated power (EIRP) is expressed in decibels. Let transmit power (P_t) is in dBW and transmit antenna gain (G) in dB, then

$$[EIRP] = [P_t] + [G] \, \mathrm{dBW} \tag{7.3}$$

Free space loss (FSL) is the major loss which occurs along the way and can be calculates as

$$L_s = \left(\frac{4\pi d}{\lambda}\right)^2 = \left(\frac{4\pi df}{C}\right)^2$$

$$[L_s] = 10 \log_{10}\left(\frac{4\pi df}{C}\right)^2 = 20 \log_{10}\left(\frac{4\pi df}{C}\right)$$

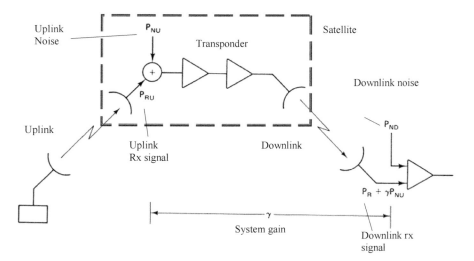

Fig. 7.9 Link budget calculation

$$[L_s] = 20 \log_{10}(4\pi) + 20 \log_{10}(d) + 20 \log_{10}(f) - 20 \log_{10}(C)$$

$$[L_s] = 20 \log_{10}(d) + 20 \log_{10}(f) - 147.55$$

$[L_s] = 20 \log_{10}(d) + 20 \log_{10}(f) + 32.45$, if d, f in kilometer and megahertz respectively.

The other losses are receiver feeder losses (RFL), antenna misalignment losses (AML), gaseous absorption, rain attenuation, cloud attenuation, polarization losses. Some of these losses are dependent on weather conditions. The attenuation effects can be mitigated using the following techniques: power control, signal processing and diversity techniques. The depolarization effects can be minimized using depolarization compensation.

System Noise

It is shown in the above section that the receiver power in a satellite link is small. Therefore, amplification is done to bring back the signal strength up to an acceptable level. This also increase the level of noise as electrical noise is present at the input. The major source of electrical noise in equipment is the thermal noise. The other noises are antenna noise, amplifier noise, intermodulation noise, intrasystem noise and so on. The noise temperature of various sources which are connected together can be added to have the total noise.

Thermal noise

As C and K bands are used for both terrestrial and satellite communication, the EIRP of the satellite systems is limited to 28 dBW in the 4 GHz band and 48 dBW in the 11 GHz band to minimize interference with terrestrial microwave communication systems. As a result, signals received by a satellite in the uplink and by an earth station in the downlink are very weak and are comparable to the thermal noise in the systems. Therefore, special considerations are required to minimize thermal noise in satellite communication systems. Thermal noise is conveniently expressed in terms of equivalent noise temperature as it is directly proportional to the temperature of the system for a given bandwidth. The thermal noise power P_n is given by

$$P_n = kT_sB$$

where, T_s = absolute temperature of the system in degrees Kelvin, B = bandwidth in Hz, K = Boltzmann's constant expressed in watts per degree Kelvin per Hz = 1.38×10^{-23} W/K/Hz.

Amplifier noise temperature

Let consider a cascaded system as shown in Fig. 7.10. For this arrangement, we will calculate the total noise energy referred to amplifier 2 input.

Fig. 7.10 Two amplifiers in cascade to find equivalent noise temperature

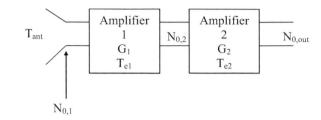

The noise output from amplifier 1 is

$$N_{0,2} = G_1 k (T_{ant} + T_{e1})$$

The noise output from amplifier 2 is

$$N_{0,out} = G_2 [G_1 k (T_{ant} + T_{e1}) + k T_{e2}]$$

The overall gain of the cascade connection is

$$G = G_1 G_2$$

Thus, the total noise energy refereed to the input of amplifier 1:

$$N_{0,1} = \frac{N_{0,out}}{G}$$

$$N_{0,1} = k \left(T_{ant} + T_{e1} + \frac{T_{e2}}{G_1} \right)$$

A System Noise temperature may now be defined as T_s by

$$N_{0,1} = k T_s$$

Thus, the overall system noise temperature referred to the input can be written as

$$T_s = T_{ant} + T_{e1} + \frac{T_{e2}}{G_1}$$

Figure 7.11 shows a typical receiving system. Based on the results obtained in the previous section, the overall noise temperature referred to the input can be expressed as

$$T_s = T_{ant} + T_{e1} + \frac{(L - 1)T_0}{G_1} + \frac{L(F - 1)T_0}{G_1} \qquad (7.4)$$

[F] is the receiver noise figure, which can be expressed in decibels.

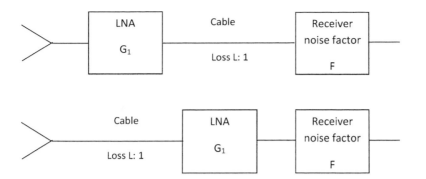

Fig. 7.11 Connections used in typical receiving system

7.6.1 *Carrier-to-Noise Ratio*

The carrier to noise ratio is the measure of the performance of a satellite link. Link budget calculation affects in determining this ratio. This ratio can be expressed as:

$$\left[\frac{C}{N}\right] = [P_R] - [P_N]\text{dB} \tag{7.5}$$

$$\left[\frac{C}{N}\right] = [EIRP] + [G_R] - [losses] - [k] - [T_s] - [B_N] \tag{7.6}$$

The $\frac{G}{T}$ ratio is a key parameter in specifying the receiving system performance. The antenna gain G_R and the system noise temperature T_s can be combined in Eqn. (7.5) as

$$[G/T] = [G_R] - [T_S] \quad \text{dBK}^{-1} \tag{7.7}$$

Therefore, the link Eq. 7.5 becomes

$$\left[\frac{C}{N}\right] = [EIRP] + [G/T] - [losses] - [k] - [B_N]. \tag{7.8}$$

7.7 Launch Vehicles and Satellite Installation

Due to the large investment involved in the launch and introduction of a major satellite system, manufacturers of spacecraft, launch vehicles, or ground systems largely deal with organizations that can make the necessary capital investment. They, in turn, must have market power in their proposed area of service. The trend in GEO satellites has been to reach as long a life as possible, extending past 10–20 years or more. If we were to move back to 10 years or even 8 years, it would be possible to reduce total propellant mass. A shorter lifetime may often have little impact on profitability of the investment because of the effect of discounting the later years of revenue. It might also be possible to increase lifetime by using a different launch vehicle to place the satellite into transfer orbit. In most missions, some of the propulsion system propellant is reserved for orbit corrections and even for the perigee kick function.

7.7.1   Launch Vehicle

The launch vehicle places the communication satellite in the desired orbit. The size and weight of the satellite to be launched depend on the launch vehicle selected. Satellite launch vehicles can be categorized into two-Expendable and Reusable.

Expendable launch vehicle is designed for single use and most of the launch vehicles are expendable. In contrast, reusable launch vehicle Space is designed for multiple time use. Most of the launches happen from the ground, however, Sea Launch has embarked on the launching of satellites from offshore platforms and Peagasus launch vehicles can launch small satellites from aircrafts. Launching of a satellite in orbit is expensive and, therefore, a number of programs have been undertaken by NASA to make the future launching of satellites in orbit as cost effective and routine as commercial air travel.

7.7.2   Satellite Installation

Installation in orbit consists of placing the satellite in its nominal orbit from a launching site on the earth surface. A launch vehicle, which could have different auxiliary propulsion systems, is used to place the satellite into an intermediate orbit called the transfer orbit. The procedure using a transfer orbit is based on the so-called Hohmann transfer which enables the satellite to move from a low altitude circular orbit to a higher altitude circular orbit with a minimum expenditure of energy. The first velocity increment changes the low altitude circular orbit into the transfer orbit which is an elliptical one whose perigee (closest point from the earth) altitude is that of the circular orbit (the velocity vector before and just after the velocity increment is perpendicular to the radius vector of the orbit), and the altitude of the apogee depends on the magnitude of the applied velocity increment. A second velocity increment at the apogee of the transfer orbit enables a circular orbit to be obtained at the altitude of the apogee (furthest point from the earth).

7.7.3   Launch of Communication Satellites

Space Shuttle is a part of lunch vehicle or carrier rocket that is used to carry communications satellites. Once the satellite in space, it is guided to be placed into orbit and to remain there. The three geostationary satellites could provide communications coverage across the whole part of the earth and are placed in orbit at an altitude of 22,300 miles above the earth. The satellites are rotated at the same speed as of the earth. As a result, the satellite appears to be stationary above the same location on earth.

7.7.4   Launch Failure

The satellite operator must keep provision for the distinct possibility that a given launch will not be successful. Spacecraft manufacturers offer a variety of services to compensate in case of failure to be placed the satellite in its specific orbit and provide services. For example, the contract for the satellite might include a provision for a second spacecraft to be ready for backup launch within a specified period after the failure.

Problem 7.1
A Satellite is orbiting in a geosynchronous orbit of radius 42,000 km. Find the velocity and time period of the orbit. Also, determine the change in velocity required if the radius of the orbit is to be reduced to 36,500 km. Assume $g_0 = 398,600.5 \ km^3/s^2$.

The gravitational coefficient, $g_0 = 398,600.5 \ km^3/s^2$

Radius of the orbit = 42,000 km

Velocity in the orbit, $v_s = \sqrt{\frac{g_0}{r_e + h}} = 3.08066 \ km/s$

Orbit period, $T_s = \frac{2\pi d^{3/2}}{\sqrt{g_0}} = 85,661.34 \ s$

For $r_e + h = 36,500 \ km$, $v_s = \sqrt{\frac{g_0}{r_e + h}} = 3.3046 \ km/s$

Increase in velocity = 3.3046 − 3.08066 = 0.224 km/s.

An antenna has a noise temperature of 35 K and is matched into a receiver that has a noise temperature of 100 K.

Calculate

(a) the noise power density and
(b) the noise power for a bandwidth of 36 MHz.

Solution

(a) $N_O = (35 + 100) \times 1.38 \times 10^{-23} = 1.863 \times 10^{-21}$ W/Hz

(b) $P_N = 1.863 \times 10^{-21} \times 36 \times 10^6 = 6.71 \times 10^{-14}$

$= 0.0671$ pW.

7.8   Sample Questions

1. A Satellite operates at downlink 4 GHz with a transmit power of 6 W and antenna gain of 48.2 dB. Calculate the EIRP in dBW.
2. The distance between a ground station and a satellite is 42,000 km. Calculate the free-space loss at a frequency of 6 GHz.
3. A television broadcast satellite receives signals with the following uplink and downlink specifications:

$$\text{Uplink}: \quad \begin{aligned} &\text{Frequency} = 6\,\text{GHz} \\ &\text{Saturation flux density} = -65.8\,\text{dBW/m}^2 \\ &\text{Input back - off} = 12\text{dB} \\ &\text{Satellite G/T} = -10.8\text{dB/K} \end{aligned}$$

$$\text{Downlink}: \quad \begin{aligned} &\text{Frequency} = 4\,\text{GHz} \\ &\text{Satellite saturation EIRP} = 25.7\,\text{dBW} \\ &\text{Output back - off} = 5\,\text{dB} \\ &\text{Free - space loss} = 196.7\,\text{dB} \\ &\text{Earth station G/T} = 42.2\,\text{dB/K} \end{aligned}$$

Calculate

   (i)   the carrier-to-noise ratio (C/N) for the uplink in dBHz,
   (ii) the C/N for the downlink in dBHz,
   (iii) the combined C/N in dB if the intermodulation ratio $(C/N)_{IM}$ and intrasystem interference ratio $(C/N)_{INT}$ are 20 and 24 dB respectively.

References

1. Space applications center, India, www.sac.gov.in
2. www.geosats.com
3. G. Maral, M. Bousquet, Z. Sun, *Satellite Communication Systems: Systems, Techniques and Technology*, 5th edn. December 2009, ISBN: 978-0-470-71458-4
4. A.K. Maini, *Varsha Agrawal Satellite Technology: Principles and Applications* (Wiley, 2014)

Index

© The Author(s) 2018   125
M. A. Matin, *Communication Systems for Electrical Engineers*,
SpringerBriefs in Electrical and Computer Engineering,
https://doi.org/10.1007/978-3-319-70129-5

Printed in the United States
By Bookmasters